Fundamentals of Differential Equations

Fundamentals of Differential Equations

Editor: Rosalind Lake

NY RESEARCH
PRESS

New York

Published by NY Research Press
118-35 Queens Blvd., Suite 400,
Forest Hills, NY 11375, USA
www.nyresearchpress.com

Fundamentals of Differential Equations
Edited by Rosalind Lake

International Standard Book Number: 978-1-63238-602-1 (Hardback)

Cataloging-in-Publication Data

Fundamentals of differential equations / edited by Rosalind Lake.
    p. cm.
Includes bibliographical references and index.
ISBN 978-1-63238-602-1
1. Differential equations. I. Lake, Rosalind.
QA371 .F86 2018
515.35--dc23

Contents

Preface

The mathematical equations which define the relationship of a function with its derivatives are known as differential equations. The varied types of differential equations include ordinary, partial, non-linear and linear differential equations. They have applications in diverse fields such as quantum mechanics, electrodynamics, economics, chemistry, etc. The book studies, analyses and upholds the pillars of differential equations and their utmost significance in modern times. Different approaches, evaluations and methodologies have also been included. In this textbook, constant effort has been made to make the understanding of the difficult concepts of this field as easy and informative as possible, for the readers.

To facilitate a deeper understanding of the contents of this book a short introduction of every chapter is written below:

Chapter 1- Differential equation is an equation which contains derivatives. The various types of differential equations are ordinary differential equations, non-linear differential equations, partial differential equations, linear differential equations, etc. This chapter is an overview of the subject matter incorporating all the major aspects of differential equations.

Chapter 2- First order partial differential equations help in understanding several problems concerned with science and technology. These equations are equations that are concerned with first derivatives. They are used in calculus of variations, in the construction of characteristic surfaces for hyperbolic partial differential equations, etc. This chapter will provide an integrated understanding of first-order partial differential equations.

Chapter 3- Second order partial differential equation studies problems like heat flow, magnetism, electricity and fluid dynamics. They can be classified into hyperbolic, parabolic or elliptic. The two examples of second-order partial differential equations cited in the chapter are Laplace equation and Poisson equation. The topics discussed in the chapter are of great importance to broaden the existing knowledge on second-order partial differential equations.

Chapter 4- The solutions that can be summed up together in particular linear combinations to form further solutions are known as linear differential equations. They can be ordinary differential equations and partial differential equations. The chapter closely examines the key concepts of linear differential equations to provide an extensive understanding of the subject.

Finally, I would like to thank the entire team involved in the inception of this book for their valuable time and contribution. This book would not have been possible without their efforts. I would also like to thank my friends and family for their constant support.

Editor

An Introduction to Differential Equations

Differential equation is an equation which contains derivatives. The various types of differential equations are ordinary differential equations, non-linear differential equations, partial differential equations, linear differential equations, etc. This chapter is an overview of the subject matter incorporating all the major aspects of differential equations.

Differential Equation

A differential equation is a mathematical equation that relates some function with its derivatives. In applications, the functions usually represent physical quantities, the derivatives represent their rates of change, and the equation defines a relationship between the two. Because such relations are extremely common, differential equations play a prominent role in many disciplines including engineering, physics, economics, and biology.

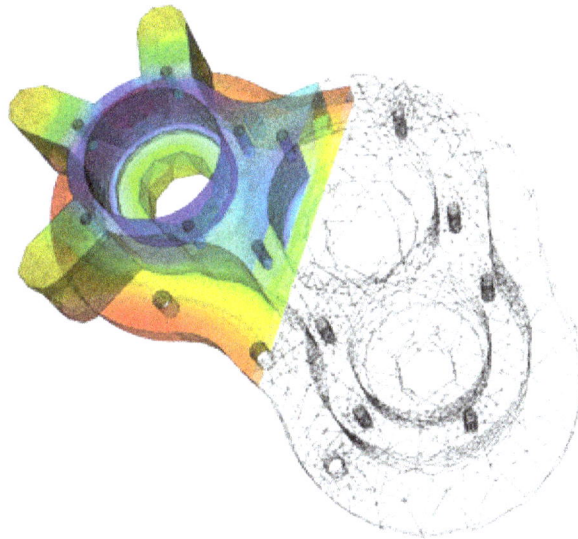

Visualization of heat transfer in a pump casing, created by solving the heat equation. Heat is being generated internally in the casing and being cooled at the boundary, providing a steady state temperature distribution.

In pure mathematics, differential equations are studied from several different perspectives, mostly concerned with their solutions—the set of functions that satisfy the equation. Only the simplest differential equations are solvable by explicit formulas; however, some properties of solutions of a given differential equation may be determined without finding their exact form.

If a self-contained formula for the solution is not available, the solution may be numerically approximated using computers. The theory of dynamical systems puts emphasis on qualitative analysis of systems described by differential equations, while many numerical methods have been developed to determine solutions with a given degree of accuracy.

History

Differential equations first came into existence with the invention of calculus by Newton and Leibniz. Isaac Newton listed three kinds of differential equations:

$$\frac{dy}{dx} = f(x)$$

$$\frac{dy}{dx} = f(x, y)$$

$$x_1 \frac{\partial y}{\partial x_1} + x_2 \frac{\partial y}{\partial x_2} = y$$

He solves these examples and others using infinite series and discusses the non-uniqueness of solutions.

Jacob Bernoulli proposed the Bernoulli differential equation in 1695. This is an ordinary differential equation of the form

$$y' + P(x)y = Q(x)y^n$$

for which the following year Leibniz obtained solutions by simplifying it.

Historically, the problem of a vibrating string such as that of a musical instrument was studied by Jean le Rond d'Alembert, Leonhard Euler, Daniel Bernoulli, and Joseph-Louis Lagrange. In 1746, d'Alembert discovered the one-dimensional wave equation, and within ten years Euler discovered the three-dimensional wave equation.

The Euler–Lagrange equation was developed in the 1750s by Euler and Lagrange in connection with their studies of the tautochrone problem. This is the problem of determining a curve on which a weighted particle will fall to a fixed point in a fixed amount of time, independent of the starting point.

Lagrange solved this problem in 1755 and sent the solution to Euler. Both further developed Lagrange's method and applied it to mechanics, which led to the formulation of Lagrangian mechanics.

Fourier published his work on heat flow in *Théorie analytique de la chaleur* (The Analytic Theory of Heat), in which he based his reasoning on Newton's law of cooling, namely, that the flow of heat between two adjacent molecules is proportional to the extremely small difference of their temperatures. Contained in this book was Fourier's proposal of his heat equation for conductive diffusion of heat. This partial differential equation is now taught to every student of mathematical physics.

Example

For example, in classical mechanics, the motion of a body is described by its position and velocity as the time value varies. Newton's laws allow (given the position, velocity, acceleration and various forces acting on the body) one to express these variables dynamically as a differential equation for the unknown position of the body as a function of time.

In some cases, this differential equation (called an equation of motion) may be solved explicitly.

An example of modelling a real world problem using differential equations is the determination of the velocity of a ball falling through the air, considering only gravity and air resistance. The ball's acceleration towards the ground is the acceleration due to gravity minus the acceleration due to air resistance. Gravity is considered constant, and air resistance may be modeled as proportional to the ball's velocity. This means that the ball's acceleration, which is a derivative of its velocity, depends on the velocity (and the velocity depends on time). Finding the velocity as a function of time involves solving a differential equation and verifying its validity.

Types

Differential equations can be divided into several types. Apart from describing the properties of the equation itself, these classes of differential equations can help inform the choice of approach to a solution. Commonly used distinctions include whether the equation is: Ordinary/Partial, Linear/Non-linear, and Homogeneous/Inhomogeneous. This list is far from exhaustive; there are many other properties and subclasses of differential equations which can be very useful in specific contexts.

Ordinary Differential Equations

An ordinary differential equation (*ODE*) is an equation containing a function of one independent variable and its derivatives. The term "*ordinary*" is used in contrast with the term partial differential equation which may be with respect to *more than* one independent variable.

Linear differential equations, which have solutions that can be added and multiplied by coefficients, are well-defined and understood, and exact closed-form solutions are obtained. By contrast, ODEs that lack additive solutions are nonlinear, and solving them is far more intricate, as one can rarely represent them by elementary functions in closed form: Instead, exact and analytic solutions of ODEs are in series or integral form. Graphical and numerical methods, applied by hand or by computer, may approximate solutions of ODEs and perhaps yield useful information, often sufficing in the absence of exact, analytic solutions.

Partial Differential Equations

A partial differential equation (*PDE*) is a differential equation that contains unknown multivariable functions and their partial derivatives. (This is in contrast to ordinary differential equations, which deal with functions of a single variable and their derivatives.) PDEs are used to formulate problems involving functions of several variables, and are either solved in closed form, or used to create a relevant computer model.

PDEs can be used to describe a wide variety of phenomena such as sound, heat, electrostatics, electrodynamics, fluid flow, elasticity, or quantum mechanics. These seemingly distinct physical phenomena can be formalised similarly in terms of PDEs. Just as ordinary differential equations often model one-dimensional dynamical systems, partial differential equations often model multidimensional systems. PDEs find their generalisation in stochastic partial differential equations.

Linear Differential Equations

A differential equation is *linear* if the unknown function and its derivatives have *degree* 1 (products of the unknown function and its derivatives are not allowed) and *nonlinear* otherwise. The characteristic property of linear equations is that their solutions form an affine subspace of an appropriate function space, which results in much more developed theory of linear differential equations.

Homogeneous linear differential equations are a subclass of linear differential equations for which the space of solutions is a linear subspace i.e. the sum of any set of solutions or multiples of solutions is also a solution. The coefficients of the unknown function and its derivatives in a linear differential equation are allowed to be (known) functions of the independent variable or variables; if these coefficients are constants then one speaks of a *constant coefficient linear differential equation*.

Non-linear Differential Equations

Non-linear differential equations are formed by the *products of the unknown function and its derivatives* are allowed and its degree is > 1. There are very few methods of solving nonlinear differential equations exactly; those that are known typically depend on the equation having particular symmetries. Nonlinear differential equations can exhibit very complicated behavior over extended time intervals, characteristic of chaos. Even the fundamental questions of existence, uniqueness, and extendability of solutions for nonlinear differential equations, and well-posedness of initial and boundary value problems for nonlinear PDEs are hard problems and their resolution in special cases is considered to be a significant advance in the mathematical theory (cf. Navier–Stokes existence and smoothness). However, if the differential equation is a correctly formulated representation of a meaningful physical process, then one expects it to have a solution.

Linear differential equations frequently appear as approximations to nonlinear equations. These approximations are only valid under restricted conditions. For example, the harmonic oscillator equation is an approximation to the nonlinear pendulum equation that is valid for small amplitude oscillations.

Equation Order

Differential equations are described by their order, determined by the term with the highest derivatives. An equation containing only first derivatives is a *first-order differential equation*, an equation containing the second derivative is a *second-order differential equation*, and so on. Differential equations that describe natural phenomena almost always have only first and second order derivatives in them, but there are some exceptions such as the thin film equation which is a fourth order partial differential equation.

Examples

In the first group of examples, let u be an unknown function of x, and let c & ω be known constants. Note both ordinary and partial differential equations are broadly classified as *linear* and *nonlinear*.

- Inhomogeneous first-order linear constant coefficient ordinary differential equation:

$$\frac{du}{dx} = cu + x^2.$$

- Homogeneous second-order linear ordinary differential equation:

$$\frac{d^2u}{dx^2} - x\frac{du}{dx} + u = 0.$$

- Homogeneous second-order linear constant coefficient ordinary differential equation describing the harmonic oscillator:

$$\frac{d^2u}{dx^2} + \omega^2 u = 0.$$

- Inhomogeneous first-order nonlinear ordinary differential equation:

$$\frac{du}{dx} = u^2 + 4.$$

- Second-order nonlinear (due to sine function) ordinary differential equation describing the motion of a pendulum of length L:

$$L\frac{d^2u}{dx^2} + g\sin u = 0.$$

In the next group of examples, the unknown function u depends on two variables x and t or x and y.

- Homogeneous first-order linear partial differential equation:

$$\frac{\partial u}{\partial t} + t\frac{\partial u}{\partial x} = 0.$$

- Homogeneous second-order linear constant coefficient partial differential equation of elliptic type, the Laplace equation:

$$\frac{\partial^2 u}{\partial x^2} + \frac{\partial^2 u}{\partial y^2} = 0.$$

$$\frac{\partial u}{\partial t} = 6u\frac{\partial u}{\partial x} - \frac{\partial^3 u}{\partial x^3}.$$

Existence of Solutions

Solving differential equations is not like solving algebraic equations. Not only are their solutions oftentimes unclear, but whether solutions are unique or exist at all are also notable subjects of interest.

For first order initial value problems, the Peano existence theorem gives one set of circumstances in which a solution exists. Given any point (a,b) in the xy-plane, define some rectangular region Z, such that $Z = [l,m] \times [n,p]$ and (a,b) is in the interior of Z. If we are given a differential equation $\frac{dy}{dx} = g(x,y)$ and the condition that $y = b$ when $x = a$, then there is locally a solution to this problem if $g(x,y)$ and $\frac{\partial g}{\partial x}$ are both continuous on Z. This solution exists on some interval with its center at a. The solution may not be unique.

However, this only helps us with first order initial value problems. Suppose we had a linear initial value problem of the nth order:

$$f_n(x)\frac{d^n y}{dx^n} + \cdots + f_1(x)\frac{dy}{dx} + f_0(x)y = g(x)$$

such that

$$y(x_0) = y_0, y'(x_0) = y_0', y''(x_0) = y_0'', \cdots$$

For any nonzero $f_n(x)$, if $\{f_0, f_1, \ \}$ and g are continuous on some interval containing x_0, y is unique and exists.

Related Concepts

- A delay differential equation (DDE) is an equation for a function of a single variable, usually called time, in which the derivative of the function at a certain time is given in terms of the values of the function at earlier times.

- A stochastic differential equation (SDE) is an equation in which the unknown quantity is a stochastic process and the equation involves some known stochastic processes, for example, the Wiener process in the case of diffusion equations.

- A differential algebraic equation (DAE) is a differential equation comprising differential and algebraic terms, given in implicit form.

Connection to Difference Equations

The theory of differential equations is closely related to the theory of difference equations, in which the coordinates assume only discrete values, and the relationship involves values of the unknown function or functions and values at nearby coordinates. Many methods to compute numerical solutions of differential equations or study the properties of differential equations involve approximation of the solution of a differential equation by the solution of a corresponding difference equation.

Applications

The study of differential equations is a wide field in pure and applied mathematics, physics, and engineering. All of these disciplines are concerned with the properties of differential equations of various types. Pure mathematics focuses on the existence and uniqueness of solutions, while applied mathematics emphasizes the rigorous justification of the methods for approximating solutions. Differential equations play an important role in modelling virtually every physical, technical, or biological process, from celestial motion, to bridge design, to interactions between neurons. Differential equations such as those used to solve real-life problems may not necessarily be directly solvable, i.e. do not have closed form solutions. Instead, solutions can be approximated using numerical methods.

Many fundamental laws of physics and chemistry can be formulated as differential equations. In biology and economics, differential equations are used to model the behavior of complex systems. The mathematical theory of differential equations first developed together with the sciences where the equations had originated and where the results found application. However, diverse problems, sometimes originating in quite distinct scientific fields, may give rise to identical differential equations. Whenever this happens, mathematical theory behind the equations can be viewed as a unifying principle behind diverse phenomena. As an example, consider propagation of light and sound in the atmosphere, and of waves on the surface of a pond. All of them may be described by the same second-order partial differential equation, the wave equation, which allows us to think of light and sound as forms of waves, much like familiar waves in the water. Conduction of heat, the theory of which was developed by Joseph Fourier, is governed by another second-order partial differential equation, the heat equation. It turns out that many diffusion processes, while seemingly different, are described by the same equation; the Black–Scholes equation in finance is, for instance, related to the heat equation.

Physics

- Euler–Lagrange equation in classical mechanics
- Hamilton's equations in classical mechanics
- Radioactive decay in nuclear physics
- Newton's law of cooling in thermodynamics
- The wave equation
- The heat equation in thermodynamics
- Laplace's equation, which defines harmonic functions
- Poisson's equation
- The geodesic equation
- The Navier–Stokes equations in fluid dynamics
- The Diffusion equation in stochastic processes

- The Convection–diffusion equation in fluid dynamics

- The Cauchy–Riemann equations in complex analysis

- The Poisson–Boltzmann equation in molecular dynamics

- The shallow water equations

- Universal differential equation

- The Lorenz equations whose solutions exhibit chaotic flow.

Classical Mechanics

So long as the force acting on a particle is known, Newton's second law is sufficient to describe the motion of a particle. Once independent relations for each force acting on a particle are available, they can be substituted into Newton's second law to obtain an ordinary differential equation, which is called the *equation of motion*.

Electrodynamics

Maxwell's equations are a set of partial differential equations that, together with the Lorentz force law, form the foundation of classical electrodynamics, classical optics, and electric circuits. These fields in turn underlie modern electrical and communications technologies. Maxwell's equations describe how electric and magnetic fields are generated and altered by each other and by charges and currents. They are named after the Scottish physicist and mathematician James Clerk Maxwell.

General Relativity

The Einstein field equations (EFE; also known as "Einstein's equations") are a set of ten partial differential equations in Albert Einstein's general theory of relativity which describe the fundamental interaction of gravitation as a result of spacetime being curved by matter and energy. First published by Einstein in 1915 as a tensor equation, the EFE equate local spacetime curvature (expressed by the Einstein tensor) with the local energy and momentum within that spacetime (expressed by the stress–energy tensor).

Quantum Mechanics

In quantum mechanics, the analogue of Newton's law is Schrödinger's equation (a partial differential equation) for a quantum system (usually atoms, molecules, and subatomic particles whether free, bound, or localized). It is not a simple algebraic equation, but in general a linear partial differential equation, describing the time-evolution of the system's wave function (also called a "state function").

Biology

- Verhulst equation – biological population growth

- von Bertalanffy model – biological individual growth

- Replicator dynamics – found in theoretical biology

- Hodgkin–Huxley model – neural action potentials

Predator-prey Equations

The Lotka–Volterra equations, also known as the predator–prey equations, are a pair of first-order, non-linear, differential equations frequently used to describe the population dynamics of two species that interact, one as a predator and the other as prey.

Chemistry

The *rate law* or rate equation for a chemical reaction is a differential equation that links the reaction rate with concentrations or pressures of reactants and constant parameters (normally rate coefficients and partial reaction orders). To determine the rate equation for a particular system one combines the reaction rate with a mass balance for the system.

Economics

- The key equation of the Solow–Swan model is $\dfrac{\partial k(t)}{\partial t} = s[k(t)]^\alpha - \delta k(t)$

- The Black–Scholes PDE

- Malthusian growth model

- The Vidale–Wolfe advertising model

Ordinary Differential Equation

In mathematics, an ordinary differential equation (ODE) is a differential equation containing one or more functions of one independent variable and its derivatives. The term *ordinary* is used in contrast with the term partial differential equation which may be with respect to *more than* one independent variable.

ODEs that are linear differential equations have exact closed-form solutions that can be added and multiplied by coefficients. By contrast, ODEs that lack additive solutions are nonlinear, and solving them is far more intricate, as one can rarely represent them by elementary functions in closed form: Instead, exact and analytic solutions of ODEs are in series or integral form. Graphical and numerical methods, applied by hand or by computer, may approximate solutions of ODEs and perhaps yield useful information, often sufficing in the absence of exact, analytic solutions.

Background

Ordinary differential equations (ODEs) arise in many contexts of mathematics and science (social as well as natural). Mathematical descriptions of change use differentials and derivatives. Various differentials, derivatives, and functions become related to each other via equations, and thus a differential equation is a result that describes dynamically changing phenomena, evolution, and variation. Often, quantities are defined as the rate of change of other quantities (for example, derivatives of displacement with respect to time), or gradients of quantities, which is how they enter differential equations.

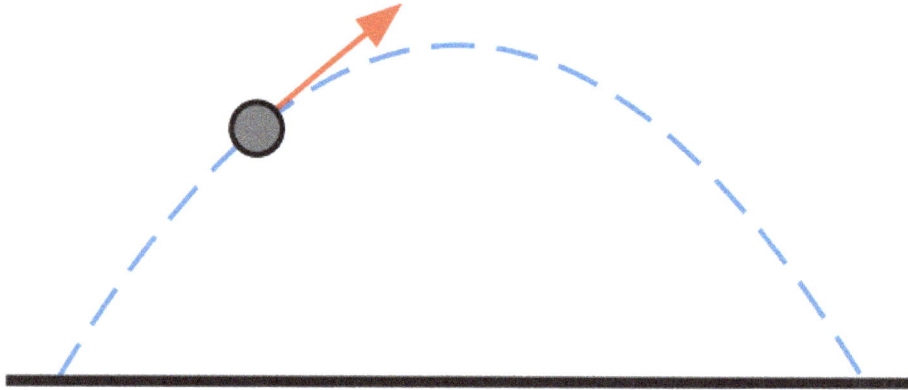

The trajectory of a projectile launched from a cannon follows a curve determined by an ordinary differential equation that is derived from Newton's second law.

Specific mathematical fields include geometry and analytical mechanics. Scientific fields include much of physics and astronomy (celestial mechanics), meteorology (weather modelling), chemistry (reaction rates), biology (infectious diseases, genetic variation), ecology and population modelling (population competition), economics (stock trends, interest rates and the market equilibrium price changes).

Many mathematicians have studied differential equations and contributed to the field, including Newton, Leibniz, the Bernoulli family, Riccati, Clairaut, d'Alembert, and Euler.

A simple example is Newton's second law of motion — the relationship between the displacement x and the time t of an object under the force F, is given by the differential equation

$$m\frac{d^2 x(t)}{dt^2} = F(x(t))$$

which constrains the motion of a particle of constant mass m. In general, F is a function of the position $x(t)$ of the particle at time t. The unknown function $x(t)$ appears on both sides of the differential equation, and is indicated in the notation $F(x(t))$.

Definitions

In what follows, let y be a dependent variable and x an independent variable, and $y = f(x)$ is an unknown function of x. The notation for differentiation varies depending upon the author and upon which notation is most useful for the task at hand. In this context, the Leibniz's notation $(dy/dx, d^2y/dx^2, ..., d^ny/dx^n)$ is more useful for differentiation and integration, whereas Lagrange's notation $(y', y'', ..., y^{(n)})$ is more useful for representing derivatives of any order compactly, and Newton's notation $(\dot{y}, \ddot{y}, \dddot{y})$ is often used in physics for representing derivatives of low order with respect to time.

General Definition

Given F, a function of x, y, and derivatives of y. Then an equation of the form

$$F\left(x, y, y', \ldots, y^{(n-1)}\right) = y^{(n)}$$

is called an explicit *ordinary differential equation* of *order n*.

More generally, an *implicit* ordinary differential equation of order n takes the form:

$$F\left(x, y, y', y'', \ldots, y^{(n)}\right) = 0$$

There are further classifications:

Autonomous

> A differential equation not depending on x is called *autonomous*.

Linear

> A differential equation is said to be *linear* if F can be written as a linear combination of the derivatives of y:

$$y^{(n)} = \sum_{i=0}^{n-1} a_i(x) y^{(i)} + r(x)$$

> where $a_i(x)$ and $r(x)$ are continuous functions in x. The function $r(x)$ is called the *source term*, leading to two further important classifications:

Homogeneous

> If $r(x) = 0$, and consequently one "automatic" solution is the trivial solution, $y = 0$. The solution of a linear homogeneous equation is a complementary function, denoted here by y_c.

Nonhomogeneous (or inhomogeneous)

> If $r(x) \neq 0$. The additional solution to the complementary function is the particular integral, denoted here by y_p.

> The general solution to a linear equation can be written as $y = y_c + y_p$.

Non-linear

> A differential equation that cannot be written in the form of a linear combination.

System of ODEs

A number of coupled differential equations form a system of equations. If y is a vector whose elements are functions; $y(x) = [y_1(x), y_2(x), \ldots, y_m(x)]$, and F is a vector-valued function of y and its derivatives, then

$$y^{(n)} = F\left(x, y, y', y'', \ldots, y^{(n-1)}\right)$$

is an *explicit system of ordinary differential equations* of *order n* and *dimension m*. In column vector form:

$$\begin{pmatrix} y_1^{(n)} \\ y_2^{(n)} \\ \vdots \\ y_m^{(n)} \end{pmatrix} = \begin{pmatrix} f_1\left(x, y, y', y'', \ldots, y^{(n-1)}\right) \\ f_2\left(x, y, y', y'', \ldots, y^{(n-1)}\right) \\ \vdots \\ f_m\left(x, y, y', y'', \ldots, y^{(n-1)}\right) \end{pmatrix}$$

These are not necessarily linear. The *implicit* analogue is:

$$F\left(x, y, y', y'', \ldots, y^{(n)}\right) = 0$$

where $0 = (0, 0, \ldots 0)$ is the zero vector. In matrix form

$$\begin{pmatrix} f_1(x, y, y', y'', \ldots, y^{(n)}) \\ f_2(x, y, y', y'', \ldots, y^{(n)}) \\ \vdots \\ f_m(x, y, y', y'', \ldots, y^{(n)}) \end{pmatrix} = \begin{pmatrix} 0 \\ 0 \\ \vdots \\ 0 \end{pmatrix}$$

For a system of the form $F(x, y, y') = 0$, some sources also require that the Jacobian matrix $\dfrac{\partial F(x, u, v)}{\partial v}$ be non-singular in order to call this an implicit ODE [system]; an implicit ODE system satisfying this Jacobian non-singularity condition can be transformed into an explicit ODE system. In the same sources, implicit ODE systems with a singular Jacobian are termed differential algebraic equations (DAEs). This distinction is not merely one of terminology; DAEs have fundamentally different characteristics and are generally more involved to solve than (nonsigular) ODE systems. Presumably for additional derivatives, the Hessian matrix and so forth are also assumed non-singular according to this scheme, although note that any ODE of order greater than one can be [and usually is] rewritten as system of ODEs of first order, which makes the Jacobian singularity criterion sufficient for this taxonomy to be comprehensive at all orders.

Solutions

Given a differential equation

$$F\left(x, y, y', \ldots, y^{(n)}\right) = 0$$

a function $u: I \subset \mathrm{R} \to \mathrm{R}$ is called a *solution* or integral curve for F, if u is n-times differentiable on I, and

$$F(x,u,u',\ldots,u^{(n)}) = 0 \quad x \in I.$$

Given two solutions $u\colon J \subset \mathrm{R} \to \mathrm{R}$ and $v\colon I \subset \mathrm{R} \to \mathrm{R}$, u is called an *extension* of v if $I \subset J$ and

$$u(x) = v(x) \quad x \in I.$$

A solution that has no extension is called a *maximal solution*. A solution defined on all of R is called a *global solution*.

A *general solution* of an nth-order equation is a solution containing n arbitrary independent constants of integration. A *particular solution* is derived from the general solution by setting the constants to particular values, often chosen to fulfill set 'initial conditions or boundary conditions'. A singular solution is a solution that cannot be obtained by assigning definite values to the arbitrary constants in the general solution.

Theories

Singular Solutions

The theory of singular solutions of ordinary and partial differential equations was a subject of research from the time of Leibniz, but only since the middle of the nineteenth century did it receive special attention. A valuable but little-known work on the subject is that of Houtain (1854). Darboux (starting in 1873) was a leader in the theory, and in the geometric interpretation of these solutions he opened a field worked by various writers, notable ones being Casorati and Cayley. To the latter is due (1872) the theory of singular solutions of differential equations of the first order as accepted circa 1900.

Reduction to Quadratures

The primitive attempt in dealing with differential equations had in view a reduction to quadratures. As it had been the hope of eighteenth-century algebraists to find a method for solving the general equation of the nth degree, so it was the hope of analysts to find a general method for integrating any differential equation. Gauss (1799) showed, however, that the differential equation meets its limitations very soon unless complex numbers are introduced. Hence, analysts began to substitute the study of functions, thus opening a new and fertile field. Cauchy was the first to appreciate the importance of this view. Thereafter, the real question was to be not whether a solution is possible by means of known functions or their integrals but whether a given differential equation suffices for the definition of a function of the independent variable or variables, and, if so, what are the characteristic properties of this function.

Fuchsian Theory

Two memoirs by Fuchs (*Crelle*, 1866, 1868), inspired a novel approach, subsequently elaborated by Thomé and Frobenius. Collet was a prominent contributor beginning in 1869, although his method for integrating a non-linear system was communicated to Bertrand in 1868. Clebsch (1873) attacked the theory along lines parallel to those followed in his theory of Abelian integrals. As the latter can be classified according to the properties of the fundamental curve that remains unchanged under a rational transformation, so Clebsch proposed to classify the transcendent functions defined by the differential equations according to the invariant properties of the corresponding surfaces $f = 0$ under rational one-to-one transformations.

Lie's Theory

From 1870, Sophus Lie's work put the theory of differential equations on a more satisfactory foundation. He showed that the integration theories of the older mathematicians can, by the introduction of what are now called Lie groups, be referred to a common source, and that ordinary differential equations that admit the same infinitesimal transformations present comparable difficulties of integration. He also emphasized the subject of transformations of contact.

Lie's group theory of differential equations has been certified, namely: (1) that it unifies the many ad hoc methods known for solving differential equations, and (2) that it provides powerful new ways to find solutions. The theory has applications to both ordinary and partial differential equations.

A general approach to solve DEs uses the symmetry property of differential equations, the continuous infinitesimal transformations of solutions to solutions (Lie theory). Continuous group theory, Lie algebras, and differential geometry are used to understand the structure of linear and nonlinear (partial) differential equations for generating integrable equations, to find its Lax pairs, recursion operators, Bäcklund transform, and finally finding exact analytic solutions to the DE.

Symmetry methods have been recognized to study differential equations, arising in mathematics, physics, engineering, and many other disciplines.

Sturm–Liouville Theory

Sturm–Liouville theory is a theory of a special type of second order ordinary differential equations. Their solutions are based on eigenvalues and corresponding eigenfunctions of linear operators defined in terms of second-order homogeneous linear equations. The problems are identified as Sturm-Liouville Problems (SLP) and are named after J.C.F. Sturm and J. Liouville, who studied such problems in the mid-1800s. The interesting fact about regular SLPs is that they have an infinite number of eigenvalues, and the corresponding eigenfunctions form a complete, orthogonal set, which makes orthogonal expansions possible. This is a key idea in applied mathematics, physics, and engineering. SLPs are also useful in the analysis of certain partial differential equations.

Existence and Uniqueness of Solutions

There are several theorems that establish existence and uniqueness of solutions to initial value problems involving ODEs both locally and globally. The two main theorems are

Theorem	Assumption	Conclusion
Peano existence theorem	F continuous	local existence only
Picard–Lindelöf theorem	F Lipschitz continuous	local existence and uniqueness

which are both local results.

Note that uniqueness theorems like the Lipschitz one above do not apply to DAE systems, which may have multiple solutions stemming from their (non-linear) algebraic part alone.

Local Existence and Uniqueness Theorem Simplified

The theorem can be stated simply as follows. For the equation and initial value problem:

$$y' = F(x, y), \quad y_0 = y(x_0)$$

if F and $\partial F/\partial y$ are continuous in a closed rectangle

$$R = [x_0 - a, x_0 + a] \times [y_0 - b, y_0 + b]$$

in the x-y plane, where a and b are real (symbolically: $a, b \in \mathbb{R}$) and $\times$ denotes the cartesian product, square brackets denote closed intervals, then there is an interval

$$I = [x_0 - h, x_0 + h] \subset [x_0 - a, x_0 + a]$$

for some $h \in \mathbb{R}$ where *the* solution to the above equation and initial value problem can be found. That is, there is a solution and it is unique. Since there is no restriction on F to be linear, this applies to non-linear equations that take the form $F(x, y)$, and it can also be applied to systems of equations.

Global Uniqueness and Maximum Domain of Solution

When the hypotheses of the Picard–Lindelöf theorem are satisfied, then local existence and uniqueness can be extended to a global result. More precisely:

For each initial condition (x_0, y_0) there exists a unique maximum (possibly infinite) open interval

$$I_{max} = (x_-, x_+), x_\pm \in \mathbb{R}, x_0 \in I_{max}$$

such that any solution that satisfies this initial condition is a restriction of the solution that satisfies this initial condition with domain I_{max}.

In the case that $x_\pm \nrightarrow \pm\infty$, there are exactly two possibilities

- explosion in finite time: $\limsup\limits_{x \to x_\pm} \| y(x) \| \to \infty$
- leaves domain of definition: $\lim\limits_{x \to x_\pm} y(x) \in \partial\overline{\Omega}$

where Ω is the open set in which F is defined, and $\partial\overline{\Omega}$ is its boundary.

Note that the maximum domain of the solution

- is always an interval (to have uniqueness)

- may be smaller than $\mathbb{R}$

- may depend on the specific choice of (x_0, y_0).

Example

$$y' = y^2$$

This means that $F(x, y) = y^2$, which is C^1 and therefore locally Lipschitz continuous, satisfying the Picard–Lindelöf theorem.

Even in such a simple setting, the maximum domain of solution cannot be all $\mathbb{R}$, since the solution is

$$y(x) = \frac{y_0}{(x_0 - x)y_0 + 1}$$

which has maximum domain:

$$
\begin{cases}
\mathbb{R} & y_0 = 0 \\
(-\infty, x_0 + \dfrac{1}{y_0}) & y_0 > 0 \\
(x_0 + \dfrac{1}{y_0}, +\infty) & y_0 < 0
\end{cases}
$$

This shows clearly that the maximum interval may depend on the initial conditions. The domain of y could be taken as being $R \setminus (x_0 + 1/y_0)$, but this would lead to a domain that is not an interval, so that the side opposite to the initial condition would be disconnected from the initial condition, and therefore not uniquely determined by it.

The maximum domain is not $\mathbb{R}$ because

$$\lim_{x \to x_{\pm}} \| y(x) \| \to \infty,$$

which is one of the two possible cases according to the above theorem.

Reduction of Order

Differential equations can usually be solved more easily if the order of the equation can be reduced.

Reduction to a First-order System

Any explicit differential equation of order n,

$$F\left(x, y, y', y'', \ldots, y^{(n-1)}\right) = y^{(n)}$$

can be written as a system of n first-order differential equations by defining a new family of unknown functions

$$y_i = y^{(i-1)}.$$

for $i = 1, 2,..., n$. The n-dimensional system of first-order coupled differential equations is then

$$
\begin{aligned}
y_1' &= y_2 \\
y_2' &= y_3 \\
&\vdots \\
y_{n-1}' &= y_n \\
y_n' &= F(x, y_1,..., y_n).
\end{aligned}
$$

more compactly in vector notation:

$$y' = F(x, y)$$

where

$$y = (y_1,..., y_n), \quad F(x, y_1,..., y_n) = (y_2,..., y_n, F(x, y_1,..., y_n)).$$

Software for ODE Solving

- Maxima computer algebra system (GPL)

- COPASI a free (Artistic License 2.0) software package for the integration and analysis of ODEs.

- MATLAB a Technical Computing Software (MATrix LABoratory)

- GNU Octave a high-level language, primarily intended for numerical computations.

- Scilab open source software for numerical computation.

- Maple

- Mathematica

- Julia (programming language)

- SciPy a Python package that includes an ODE integration module.

- Chebfun an open-source package, written in MATLAB, for computing with functions to 15-digit accuracy.

- GNU R an open source computational environment primarily intended for statistics, which includes package for ODE solving.

- EROS.NET a free ODE solver for .NET.

Partial Differential Equation

In mathematics, a partial differential equation (PDE) is a differential equation that contains unknown multivariable functions and their partial derivatives. (A special case are ordinary differential equations (ODEs), which deal with functions of a single variable and their derivatives.) PDEs are used to formulate problems involving functions of several variables, and are either solved by hand, or used to create a relevant computer model.

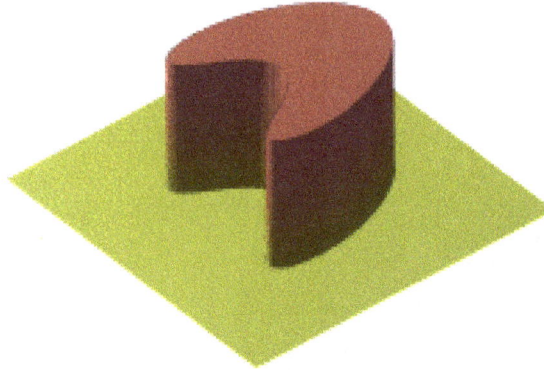

A visualisation of a solution to the two-dimensional heat equation with
temperature represented by the third dimension

PDEs can be used to describe a wide variety of phenomena such as sound, heat, electrostatics, electrodynamics, fluid dynamics, elasticity, or quantum mechanics. These seemingly distinct physical phenomena can be formalised similarly in terms of PDEs. Just as ordinary differential equations often model one-dimensional dynamical systems, partial differential equations often model multidimensional systems. PDEs find their generalisation in stochastic partial differential equations.

Partial differential equations (PDEs) are equations that involve rates of change with respect to continuous variables. The position of a rigid body is specified by six numbers, but the configuration of a fluid is given by the continuous distribution of several parameters, such as the temperature, pressure, and so forth. The dynamics for the rigid body take place in a finite-dimensional configuration space; the dynamics for the fluid occur in an infinite-dimensional configuration space. This distinction usually makes PDEs much harder to solve than ordinary differential equations (ODEs), but here again, there will be simple solutions for linear problems. Classic domains where PDEs are used include acoustics, fluid dynamics, electrodynamics, and heat transfer.

A partial differential equation (PDE) for the function $u(x_1, \cdots, x_n)$ is an equation of the form

$$f\left(x_1, \ldots, x_n, u, \frac{\partial u}{\partial x_1}, \ldots, \frac{\partial u}{\partial x_n}, \frac{\partial^2 u}{\partial x_1 \partial x_1}, \ldots, \frac{\partial^2 u}{\partial x_1 \partial x_n}, \ldots \right) = 0.$$

If f is a linear function of u and its derivatives, then the PDE is called linear. Common examples of linear PDEs include the heat equation, the wave equation, Laplace's equation, Helmholtz equation, Klein–Gordon equation, and Poisson's equation.

A relatively simple PDE is

$$\frac{\partial u}{\partial x}(x, y) = 0.$$

This relation implies that the function $u(x,y)$ is independent of x. However, the equation gives no information on the function's dependence on the variable y. Hence the general solution of this equation is

$$u(x, y) = f(y),$$

where f is an arbitrary function of y. The analogous ordinary differential equation is

$$\frac{du}{dx}(x) = 0,$$

which has the solution

$$u(x) = c,$$

where c is any constant value. These two examples illustrate that general solutions of ordinary differential equations (ODEs) involve arbitrary constants, but solutions of PDEs involve arbitrary functions. A solution of a PDE is generally not unique; additional conditions must generally be specified on the boundary of the region where the solution is defined. For instance, in the simple example above, the function $f(y)$ can be determined if u is specified on the line $x = 0$.

Existence and Uniqueness

Although the issue of existence and uniqueness of solutions of ordinary differential equations has a very satisfactory answer with the Picard–Lindelöf theorem, that is far from the case for partial differential equations. The Cauchy–Kowalevski theorem states that the Cauchy problem for any partial differential equation whose coefficients are analytic in the unknown function and its derivatives, has a locally unique analytic solution. Although this result might appear to settle the existence and uniqueness of solutions, there are examples of linear partial differential equations whose coefficients have derivatives of all orders (which are nevertheless not analytic) but which have no solutions at all. Even if the solution of a partial differential equation exists and is unique, it may nevertheless have undesirable properties. The mathematical study of these questions is usually in the more powerful context of weak solutions.

An example of pathological behavior is the sequence (depending upon n) of Cauchy problems for the Laplace equation

$$\frac{\partial^2 u}{\partial x^2} + \frac{\partial^2 u}{\partial y^2} = 0,$$

with boundary conditions

$$u(x, 0) = 0,$$

$$\frac{\partial u}{\partial y}(x, 0) = \frac{\sin(nx)}{n},$$

where n is an integer. The derivative of u with respect to y approaches 0 uniformly in x as n increases, but the solution is

$$u(x, y) = \frac{\sinh(ny)\sin(nx)}{n^2}.$$

This solution approaches infinity if nx is not an integer multiple of π for any non-zero value of y. The Cauchy problem for the Laplace equation is called *ill-posed* or *not well-posed*, since the solution does not continuously depend on the data of the problem. Such ill-posed problems are not usually satisfactory for physical applications.

Notation

In PDEs, it is common to denote partial derivatives using subscripts. That is:

$$u_x = \frac{\partial u}{\partial x}$$

$$u_{xx} = \frac{\partial^2 u}{\partial x^2}$$

$$u_{xy} = \frac{\partial^2 u}{\partial y \partial x} = \frac{\partial}{\partial y}\left(\frac{\partial u}{\partial x}\right).$$

Especially in physics, del or Nabla (∇) is often used to denote spatial derivatives, and $\ddot{u}$ for time derivatives. For example, the wave equation (described below) can be written as

$$\ddot{u} = c^2 \nabla^2 u$$

or

$$\ddot{u} = c^2 \Delta u$$

where Δ is the Laplace operator.

Classification

Some linear, second-order partial differential equations can be classified as parabolic, hyperbolic and elliptic. Others such as the Euler–Tricomi equation have different types in different regions. The classification provides a guide to appropriate initial and boundary conditions, and to the smoothness of the solutions.

Equations of First Order

Linear Equations of Second Order

Assuming $u_{xy} = u_{yx}$, the general second-order PDE in two independent variables has the form

$$Au_{xx} + 2Bu_{xy} + Cu_{yy} + \cdots (\text{lower order terms}) = 0,$$

where the coefficients A, B, C etc. may depend upon x and y. If $A^2 + B^2 + C^2 > 0$ over a region of the xy plane, the PDE is second-order in that region. This form is analogous to the equation for a conic section:

$$Ax^2 + 2Bxy + Cy^2 + \cdots = 0.$$

More precisely, replacing ∂_x by X, and likewise for other variables (formally this is done by a Fourier transform), converts a constant-coefficient PDE into a polynomial of the same degree, with the top degree (a homogeneous polynomial, here a quadratic form) being most significant for the classification.

Just as one classifies conic sections and quadratic forms into parabolic, hyperbolic, and elliptic based on the discriminant $B^2 - 4AC$, the same can be done for a second-order PDE at a given point. However, the discriminant in a PDE is given by $B^2 - AC$, due to the convention of the xy term being $2B$ rather than B; formally, the discriminant (of the associated quadratic form) is $(2B)^2 - 4AC = 4(B^2 - AC)$, with the factor of 4 dropped for simplicity.

1. $B^2 - AC < 0$: Solutions of elliptic PDEs are as smooth as the coefficients allow, within the interior of the region where the equation and solutions are defined. For example, solutions of Laplace's equation are analytic within the domain where they are defined, but solutions may assume boundary values that are not smooth. The motion of a fluid at subsonic speeds can be approximated with elliptic PDEs, and the Euler–Tricomi equation is elliptic where $x < 0$.

2. $B^2 - AC = 0$: Equations that are parabolic at every point can be transformed into a form analogous to the heat equation by a change of independent variables. Solutions smooth out as the transformed time variable increases. The Euler–Tricomi equation has parabolic type on the line where $x = 0$.

3. $B^2 - AC > 0$: Hyperbolic equations retain any discontinuities of functions or derivatives in the initial data. An example is the wave equation. The motion of a fluid at supersonic speeds can be approximated with hyperbolic PDEs, and the Euler–Tricomi equation is hyperbolic where $x > 0$.

If there are n independent variables $x_1, x_2, \ldots, x_n$, a general linear partial differential equation of second order has the form

$$Lu = \sum_{i=1}^{n} \sum_{j=1}^{n} a_{i,j} \frac{\partial^2 u}{\partial x_i \partial x_j} \quad \text{plus lower-order terms} = 0.$$

The classification depends upon the signature of the eigenvalues of the coefficient matrix $a_{i,j}$.

1. Elliptic: The eigenvalues are all positive or all negative.

2. Parabolic: The eigenvalues are all positive or all negative, save one that is zero.

3. Hyperbolic: There is only one negative eigenvalue and all the rest are positive, or there is only one positive eigenvalue and all the rest are negative.

4. Ultrahyperbolic: There is more than one positive eigenvalue and more than one negative eigenvalue, and there are no zero eigenvalues. There is only a limited theory for ultra-hyperbolic equations (Courant and Hilbert, 1962).

Systems of first-order Equations and Characteristic Surfaces

The classification of partial differential equations can be extended to systems of first-order equations, where the unknown u is now a vector with m components, and the coefficient matrices A_v are m by m matrices for $v = 1, ..., n$. The partial differential equation takes the form

$$Lu = \sum_{v=1}^{n} A_v \frac{\partial u}{\partial x_v} + B = 0,$$

where the coefficient matrices A_v and the vector B may depend upon x and u. If a hypersurface S is given in the implicit form

$$\varphi(x_1, x_2, \ldots, x_n) = 0,$$

where φ has a non-zero gradient, then S is a characteristic surface for the operator L at a given point if the characteristic form vanishes:

$$Q\left(\frac{\partial \varphi}{\partial x_1}, \ldots, \frac{\partial \varphi}{\partial x_n}\right) = \det\left[\sum_{v=1}^{n} A_v \frac{\partial \varphi}{\partial x_v}\right] = 0.$$

The geometric interpretation of this condition is as follows: if data for u are prescribed on the surface S, then it may be possible to determine the normal derivative of u on S from the differential equation. If the data on S and the differential equation determine the normal derivative of u on S, then S is non-characteristic. If the data on S and the differential equation *do not* determine the normal derivative of u on S, then the surface is characteristic, and the differential equation restricts the data on S: the differential equation is *internal* to S.

1. A first-order system $Lu=0$ is *elliptic* if no surface is characteristic for L: the values of u on S and the differential equation always determine the normal derivative of u on S.

2. A first-order system is *hyperbolic* at a point if there is a space-like surface S with normal ξ at that point. This means that, given any non-trivial vector η orthogonal to ξ, and a scalar multiplier λ, the equation $Q(\lambda\xi + \eta) = 0$ has m real roots $\lambda_1, \lambda_2, ..., \lambda_m$. The system is strictly hyperbolic if these roots are always distinct. The geometrical interpretation of this condition is as follows: the characteristic form $Q(\zeta) = 0$ defines a cone (the normal cone) with homogeneous coordinates ζ. In the hyperbolic case, this cone has m sheets, and the axis $\zeta = \lambda \xi$ runs inside these sheets: it does not intersect any of them. But when displaced from the origin by η, this axis intersects every sheet. In the elliptic case, the normal cone has no real sheets.

Equations of Mixed Type

If a PDE has coefficients that are not constant, it is possible that it will not belong to any of these categories but rather be of mixed type. A simple but important example is the Euler–Tricomi equation

$$u_{xx} = x u_{yy},$$

which is called elliptic-hyperbolic because it is elliptic in the region $x < 0$, hyperbolic in the region $x > 0$, and degenerate parabolic on the line $x = 0$.

Infinite-order PDEs in Quantum Mechanics

In the phase space formulation of quantum mechanics, one may consider the quantum Hamilton's equations for trajectories of quantum particles. These equations are infinite-order PDEs. However, in the semiclassical expansion, one has a finite system of ODEs at any fixed order of $\hbar$. The evolution equation of the Wigner function is also an infinite-order PDE. The quantum trajectories are quantum characteristics, with the use of which one could calculate the evolution of the Wigner function.

Analytical Solutions

Separation of Variables

Linear PDEs can be reduced to systems of ordinary differential equations by the important technique of separation of variables. This technique rests on a characteristic of solutions to differential equations: if one can find any solution that solves the equation and satisfies the boundary conditions, then it is *the* solution (this also applies to ODEs). We assume as an ansatz that the dependence of a solution on the parameters space and time can be written as a product of terms that each depend on a single parameter.

In the method of separation of variables, one reduces a PDE to a PDE in fewer variables, which is an ordinary differential equation if in one variable – these are in turn easier to solve.

This is possible for simple PDEs, which are called separable partial differential equations, and the domain is generally a rectangle (a product of intervals). Separable PDEs correspond to diagonal matrices – thinking of "the value for fixed x" as a coordinate, each coordinate can be understood separately.

This generalizes to the method of characteristics, and is also used in integral transforms.

Method of Characteristics

In special cases, one can find characteristic curves on which the equation reduces to an ODE – changing coordinates in the domain to straighten these curves allows separation of variables, and is called the method of characteristics.

More generally, one may find characteristic surfaces.

Integral Transform

An integral transform may transform the PDE to a simpler one, in particular, a separable PDE. This corresponds to diagonalizing an operator.

An important example of this is Fourier analysis, which diagonalizes the heat equation using the eigenbasis of sinusoidal waves.

If the domain is finite or periodic, an infinite sum of solutions such as a Fourier series is appropriate, but an integral of solutions such as a Fourier integral is generally required for infinite domains. The solution for a point source for the heat equation given above is an example of the use of a Fourier integral.

Change of Variables

Often a PDE can be reduced to a simpler form with a known solution by a suitable change of variables. For example, the Black–Scholes PDE

$$\frac{\partial V}{\partial t} + \frac{1}{2}\sigma^2 S^2 \frac{\partial^2 V}{\partial S^2} + rS\frac{\partial V}{\partial S} - rV = 0$$

is reducible to the heat equation

$$\frac{\partial u}{\partial \tau} = \frac{\partial^2 u}{\partial x^2}$$

by the change of variables

$$V(S,t) = Kv(x,\tau)$$

$$x = \ln\left(\frac{S}{K}\right)$$

$$\tau = \tfrac{1}{2}\sigma^2(T-t)$$

$$v(x,\tau) = \exp(-\alpha x - \beta\tau)u(x,\tau).$$

Fundamental Solution

Inhomogeneous equations can often be solved (for constant coefficient PDEs, always be solved) by finding the fundamental solution (the solution for a point source), then taking the convolution with the boundary conditions to get the solution.

This is analogous in signal processing to understanding a filter by its impulse response.

Superposition Principle

Because any superposition of solutions of a linear, homogeneous PDE is again a solution, the particular solutions may then be combined to obtain more general solutions. if u_1 and u_2 are solutions of a homogeneous linear pde in same region R, then $u = c_1 u_1 + c_2 u_2$ with any constants c_1 and c_2 are also a solution of that pde in that same region....

Methods for non-linear Equations

There are no generally applicable methods to solve nonlinear PDEs. Still, existence and uniqueness results (such as the Cauchy–Kowalevski theorem) are often possible, as are proofs of important qualitative and quantitative properties of solutions (getting these results is a major part of analysis). Computational solution to the nonlinear PDEs, the split-step method, exist for specific equations like nonlinear Schrödinger equation.

Nevertheless, some techniques can be used for several types of equations. The h-principle is the most powerful method to solve underdetermined equations. The Riquier–Janet theory is an effective method for obtaining information about many analytic overdetermined systems.

The method of characteristics (similarity transformation method) can be used in some very special cases to solve partial differential equations.

In some cases, a PDE can be solved via perturbation analysis in which the solution is considered to be a correction to an equation with a known solution. Alternatives are numerical analysis techniques from simple finite difference schemes to the more mature multigrid and finite element methods. Many interesting problems in science and engineering are solved in this way using computers, sometimes high performance supercomputers.

Lie Group Method

From 1870 Sophus Lie's work put the theory of differential equations on a more satisfactory foundation. He showed that the integration theories of the older mathematicians can, by the introduction of what are now called Lie groups, be referred to a common source; and that ordinary differential equations which admit the same infinitesimal transformations present comparable difficulties of integration. He also emphasized the subject of transformations of contact.

A general approach to solving PDE's uses the symmetry property of differential equations, the continuous infinitesimal transformations of solutions to solutions (Lie theory). Continuous group theory, Lie algebras and differential geometry are used to understand the structure of linear and nonlinear partial differential equations for generating integrable equations, to find its Lax pairs, recursion operators, Bäcklund transform and finally finding exact analytic solutions to the PDE.

Symmetry methods have been recognized to study differential equations arising in mathematics, physics, engineering, and many other disciplines.

Semianalytical Methods

The adomian decomposition method, the Lyapunov artificial small parameter method, and He's homotopy perturbation method are all special cases of the more general homotopy analysis method. These are series expansion methods, and except for the Lyapunov method, are independent of small physical parameters as compared to the well known perturbation theory, thus giving these methods greater flexibility and solution generality.

Numerical Solutions

The three most widely used numerical methods to solve PDEs are the finite element method (FEM), finite volume methods (FVM) and finite difference methods (FDM). The FEM has a prominent position among these methods and especially its exceptionally efficient higher-order version hp-FEM. Other versions of FEM include the generalized finite element method (GFEM), extended finite element method (XFEM), spectral finite element method (SFEM), meshfree finite element method, discontinuous Galerkin finite element method (DGFEM), Element-Free Galerkin Method (EFGM), Interpolating Element-Free Galerkin Method (IEFGM), etc.

Finite Element Method

The finite element method (FEM) (its practical application often known as finite element analysis (FEA)) is a numerical technique for finding approximate solutions of partial differential equations (PDE) as well as of integral equations. The solution approach is based either on eliminating the differential equation completely (steady state problems), or rendering the PDE into an approximating system of ordinary differential equations, which are then numerically integrated using standard techniques such as Euler's method, Runge–Kutta, etc.

Finite Difference Method

Finite-difference methods are numerical methods for approximating the solutions to differential equations using finite difference equations to approximate derivatives.

Finite Volume Method

Similar to the finite difference method or finite element method, values are calculated at discrete places on a meshed geometry. "Finite volume" refers to the small volume surrounding each node point on a mesh. In the finite volume method, surface integrals in a partial differential equation that contain a divergence term are converted to volume integrals, using the divergence theorem. These terms are then evaluated as fluxes at the surfaces of each finite volume. Because the flux entering a given volume is identical to that leaving the adjacent volume, these methods are conservative.

Multivariable Calculus

Multivariable calculus (also known as multivariate calculus) is the extension of calculus in one variable to calculus with functions of several variables: the differentiation and integration of functions involving multiple variables, rather than just one.

Typical Operations

Limits and Continuity

A study of limits and continuity in multivariable calculus yields many counter-intuitive results not demonstrated by single-variable functions. For example, there are scalar functions of two variables with points in their domain which give a particular limit when approached along any arbitrary line, yet give a different limit when approached along a parabola.

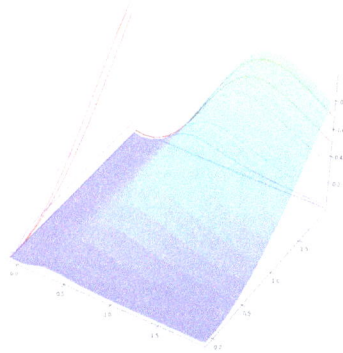

visualisation of the curve described above with projections on the planes xz and yz
(the blue lined projections are the closest to zero)

Indeed, the function

$$f(x,y) = \frac{x^2 y}{x^4 + y^2}$$

approaches zero along any line through the origin. However, when the origin is approached along a parabola $y = x^2$, it has a limit of 0.5. Since taking different paths toward the same point yields different values for the limit, the limit does not exist.

Continuity in each argument is not sufficient for multivariate continuity: For instance, in the case of a real-valued function with two real-valued parameters, $f(x,y)$, continuity of f in x for fixed y and continuity of f in y for fixed x does not imply continuity of f.

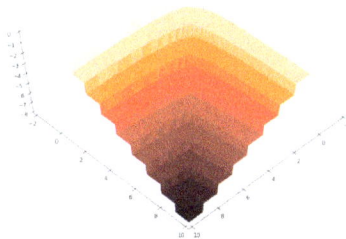

Consider

$$f(x,y) = \begin{cases} \dfrac{y}{x} - y & \text{if } 1 \geq x > y \geq 0 \\[2mm] \dfrac{x}{y} - x & \text{if } 1 \geq y > x \geq 0 \\[2mm] 1 - x & \text{if } x = y > 0 \\[2mm] 0 & \text{else.} \end{cases}$$

It is easy to verify that all real-valued functions (with one real-valued argument) that are given by $f_y(x) := f(x,y)$ are continuous in x (for any fixed y). Similarly, all f_x are continuous as f is symmetric with regards to x and y. However, f itself is not continuous as can be seen by considering the sequence $f\left(\dfrac{1}{n},\dfrac{1}{n}\right)$ (for natural n) which should converge to $f(0,0)=0$ if f was continuous. However, $\lim\limits_{n\to\infty} f\left(\dfrac{1}{n},\dfrac{1}{n}\right)=1$. Thus, function is not continuous at $(0,0)$.

Partial Differentiation

The partial derivative generalizes the notion of the derivative to higher dimensions. A partial derivative of a multivariable function is a derivative with respect to one variable with all other variables held constant.

Partial derivatives may be combined in interesting ways to create more complicated expressions of the derivative. In vector calculus, the del operator (∇) is used to define the concepts of gradient, divergence, and curl in terms of partial derivatives. A matrix of partial derivatives, the Jacobian matrix, may be used to represent the derivative of a function between two spaces of arbitrary dimension. The derivative can thus be understood as a linear transformation which directly varies from point to point in the domain of the function.

Differential equations containing partial derivatives are called partial differential equations or PDEs. These equations are generally more difficult to solve than ordinary differential equations, which contain derivatives with respect to only one variable.

Multiple Integration

The multiple integral expands the concept of the integral to functions of any number of variables. Double and triple integrals may be used to calculate areas and volumes of regions in the plane and in space. Fubini's theorem guarantees that a multiple integral may be evaluated as a *repeated integral* or *iterated integral* as long as the integrand is continuous throughout the domain of integration.

The surface integral and the line integral are used to integrate over curved manifolds such as surfaces and curves.

Fundamental Theorem of Calculus in Multiple Dimensions

In single-variable calculus, the fundamental theorem of calculus establishes a link between the derivative and the integral. The link between the derivative and the integral in multivariable calculus is embodied by the integral theorems of vector calculus:

- Gradient theorem
- Stokes' theorem
- Divergence theorem
- Green's theorem.

In a more advanced study of multivariable calculus, it is seen that these four theorems are specific incarnations of a more general theorem, the generalized Stokes' theorem, which applies to the integration of differential forms over manifolds.

Applications and Uses

Techniques of multivariable calculus are used to study many objects of interest in the material world. In particular,

		Domain/Codomain	Applicable techniques
Curves		$f : \mathbb{R} \to \mathbb{R}^n$	Lengths of curves, line integrals, and curvature.
Surfaces		$f : \mathbb{R}^2 \to \mathbb{R}^n$	Areas of surfaces, surface integrals, flux through surfaces, and curvature.
Scalar fields		$f : \mathbb{R}^n \to \mathbb{R}$	Maxima and minima, Lagrange multipliers, directional derivatives.
Vector fields		$f : \mathbb{R}^m \to \mathbb{R}^n$	Any of the operations of vector calculus including gradient, divergence, and curl.

Multivariable calculus can be applied to analyze deterministic systems that have multiple degrees of freedom. Functions with independent variables corresponding to each of the degrees of freedom are often used to model these systems, and multivariable calculus provides tools for characterizing the system dynamics.

Multivariate calculus is used in the optimal control of continuous time dynamic systems. It is used in regression analysis to derive formulas for estimating relationships among various sets of empirical data.

Multivariable calculus is used in many fields of natural and social science and engineering to model and study high-dimensional systems that exhibit deterministic behavior. In economics, for example, consumer choice over a variety of goods, and producer choice over various inputs to use and outputs to produce, are modeled with multivariate calculus. Quantitative analysts in finance also often use multivariate calculus to predict future trends in the stock market.

Non-deterministic, or stochastic systems can be studied using a different kind of mathematics, such as stochastic calculus.

We recall some basic concepts from multivariable calculus. The concepts of limits, continuity, partial derivatives, directional derivatives, chain rules, tangent plane and normals are discussed.

For any $(x, y), (x_0, y_0) \in \mathbb{R}^2$, let us denote

$$d((x, y), (x_0, y_0)) = \sqrt{(x - x_0)^2 + (y - y_0)^2}$$

for the distance between two points (x, y) and (x_0, y_0). A disk $D_r(x_0, y_0)$ of radius r centered at (x_0, y_0) is defined as

$$D_r(x_0, y_0) = \{(x, y) | d((x, y), (x_0, y_0)) < r\}.$$

The concept of limit now can be defined by the same ϵ, δ technique as in one variable calculus.

DEFINITION 1. (The ϵ, δ definition of limit) Let f (x, y) be a real-valued function of two variables defined on a disk $D_r(x_0, y_0)$, except possibly at (x_0, y_0). Then

$$\lim_{(x,y)\to(x_0,y_0)} f(x, y) = l \ \ if \ for \ every \ \epsilon > 0 \ there \ is \ a \ \delta > 0 \ \ such \ that$$

$$\left| f(x, y) - l \right| < \epsilon \ \ whenever \ 0 < d((x, y), (x_0, y_0)) < \delta$$

Definition 1 means that the distance between $f(x, y)$ and l can be made arbitrarily small by making the distance from (x, y) to (x_0, y_0) sufficiently small (but not 0). That is, if any small interval $(1 - \epsilon, 1 + \epsilon)$ is given around l, then we can find a disk $D_\delta(x_0, y_0)$ with center (x_0, y_0) and radius $\delta > 0$ such that f maps all the points in $D_\delta(x_0, y_0)$ [except possibly (x_0, y_0)] into the interval $(1 - \epsilon, 1 + \epsilon)$.

The definition of a limit can be extended to functions of three or more variables. Using vector notation the definition can be written in a compact form as follows:

Let $f : D(x) \subset \mathbb{R} \to \mathbb{R}$. Then

$$\lim_{x \to x_0} f(x) = l \ \ if \ for \ every \ \epsilon > 0 \ there \ is \ \delta > 0 \ such \ that$$

$$\left| f(x) - l \right| < \epsilon \ \ whenever \ 0 < d(x, x_0) < \delta.$$

DEFINITION 2. (Continuity) Let $f(x, y)$ be a real-valued function of two variables defined in a disk $D_r(x_0, y_0)$ with center (x_0, y_0). Then

$$f \ \ is \ \ continous \ at (x_0, y_0) \ if \lim_{(x,y)\to(x_0,y_0)} f(x, y) = f(x_0, y_0)$$

We say f is continuous in $D_r(x_0, y_0)$ if f is continuous at every point (x, y) in $D_r(x_0, y_0)$. The intuitive meaning of continuity is that if the point (x, y) changes by a small amount, then the value of $f(x, y)$ changes by a small amount. Geometrically, this means that a surface that is the graph of a continuous function has no holes or breaks.

DEFINITION 3. (Partial derivatives) *Let* $f : D_r(x_0, y_0) \to \mathbb{R}$. The partial derivatives of f are the functions f_x and f_y defined by

$$f_x(x, y) := \lim_{h \to 0} \frac{f(x+h, y) - f(x, y)}{h},$$

$$f_y(x, y) := \lim_{h \to 0} \frac{f(x, h+y) - f(x, y)}{h}.$$

To find f_x, treat y as a constant and differentiate $f(x, y)$ with respect to x. Similarly, to find f_y, treat x as a constant and differentiate $f(x, y)$ with respect to y. If $z = f(x, y)$ we write

$$f_x = \frac{\partial f}{\partial x} = \frac{\partial z}{\partial x} = z_x,$$

$$f_y = \frac{\partial f}{\partial y} = \frac{\partial z}{\partial y} = z_y,$$

Partial derivatives can also be defined for functions of three or more variables. In general, if z is a function of n variables, $z = f(x_1, x_2, \ldots\ldots, x_n)$,its partial derivative with respect to the ith variable x_i is

$$\frac{\partial z}{\partial x_i} := \lim_{h \to 0} \frac{f(x_1, \ldots, x_{i-1}, x_i + h, x_{i+1}, \ldots\ldots, x_n) - f(x_1, \ldots\ldots, x_i, \ldots\ldots, x_n)}{h}.$$

We also write

$$z_{x_i} = \frac{\partial z}{\partial x_i} = \frac{\partial f}{\partial x_i} = f_{x_i}.$$

Since the partial derivatives are themselves functions, we can take their partial derivatives to obtain higher derivatives. If $z = f(x, y)$, we may compute

$$f_{xx}(x, y) = \frac{\partial}{\partial x}\left(\frac{\partial z}{\partial x}\right) = \frac{\partial^2 z}{\partial x^2}, f_{yy}(x, y) = \frac{\partial}{\partial y}\left(\frac{\partial z}{\partial y}\right) = \frac{\partial^2 z}{\partial y^2},$$

$$f_{xy}(x, y) = \frac{\partial}{\partial y}\left(\frac{\partial z}{\partial x}\right) = \frac{\partial^2 z}{\partial y \partial x}, f_{yx}(x, y) = \frac{\partial}{\partial x}\left(\frac{\partial z}{\partial y}\right) = \frac{\partial^2 z}{\partial x \partial y}.$$

In general, $f_{xy} \neq f_{yx}$. However, the following theorem gives condition under which we can assert that $f_{xy} = f_{yx}$.

THEOREM: *Let* $f : D_r(x_0, y_0) \to \mathbb{R}$. *If* f_{xy} *and* f_{yx} *are both continuous at* (x_0, y_0), *then*

$$f_{xy}(x_0, y_0) = f_{yx}(x_0, y_0).$$

DEFINITION 5. (Chain rule) *Let* $z_1 = f_1(x_1, \ldots, x_n), \ldots, z_m = f_m(x_1, \ldots, x_n)$ be m functions of n variables, and let $x_1 = g_1(t_1, \ldots, t_k), \ldots, x_n = g_n(t_1, \ldots, t_k)$ be n functions of k variables, all with continuous partial derivatives.

Consider the $z_i's$ as functions of the $t_j's$ by

$$z_i = f_i(g_1(t_1, \ldots, t_k), \ldots, g_n(t_1, \ldots, t_k)).$$

Then

$$\frac{\partial z_i}{\partial t_j} = \frac{\partial z_i}{\partial x_1}\frac{\partial x_1}{\partial t_j} + \frac{\partial z_i}{\partial x_2}\frac{\partial x_2}{\partial t_j} + \ldots + \frac{\partial z_i}{\partial x_n}\frac{\partial x_n}{\partial t_j}.$$

DEFINITION 6. If $z = f(x, y)$ is a function of two variables, its gradient vector field ∇f is defined by

$$\nabla f(x, y) := (f_x(x, y), f_y(x, y)) = (\frac{\partial z}{\partial x}, \frac{\partial z}{\partial y}).$$

If $u = f(x, y, z)$ is a function of three variables, its gradient vector field ∇f is defined by

$$\nabla f(x, y, z) = (f_x(x, y, z), f_y(x, y, z), f_z(x, y, z)) = (\frac{\partial u}{\partial x}, \frac{\partial u}{\partial y}, \frac{\partial u}{\partial z}).$$

DEFINITION 7. (Implicit differentiation) If $y = f(x)$ is a function satisfying the relation $z = F(x, y) = 0$, then

$$\frac{dy}{dx} = -\frac{F_x(x, f(x))}{F_y(x, f(x))}.$$

Differentiating $F(x, y) = 0$ with respect to x using the chain rule gives

$$\frac{\partial F}{\partial x}\frac{dx}{dx} + \frac{\partial F}{\partial y}\frac{dy}{dx} = 0$$

$$\Rightarrow \frac{\partial F}{\partial x} + \frac{\partial F}{\partial y}\frac{dy}{dx} = 0,$$

which yields $\frac{dy}{dx} = -\frac{F_x(x, f(x))}{F_y(x, f(x))}$.

DEFINITION 8. (Directional derivatives) The directional derivative of F at (x_0, y_0) in the direction of a unit vector $u = (u_1, u_2)$ is

$$\mathbb{D}_{\mathbf{u}} f(x_0, y_0) := \lim_{h \to 0} \frac{f(x_0 + hu_1, y_0 + hu_2) - f(x_0, y_0)}{h}$$

if this limit exists.

Note that if u = (1, 0) then $\mathbb{D}_u f = f_x$ and if u = (0, 1), then $\mathbb{D}_u f = f_y$. In other words, the partial derivatives of f with respect to x and y are just special cases of the directional derivatives.

THEOREM: If $f(x, y)$ is a differentiable function of x and y, then f has a directional derivative in the direction of any unit vector $u = (u_1, u_2)$ and

$$\mathbb{D}_{\mathbf{u}} f(x, y) = f_x(x, y) u_1 + f_y(x, y) u_2.$$

The directional derivative can be written as

$$\begin{aligned} \mathbb{D}_{\mathbf{u}} f(x, y) &= f_x(x, y) u_1 + f_y(x, y) u_2 \\ &= (f_x(x, y), f_y(x, y)) \cdot (u_1, u_2) \\ &= \nabla f(x, y) \cdot \mathbf{u}. \end{aligned}$$

This expresses the directional derivative in the direction of u as the scalar projection of the gradient vector onto u. From above equation, we have

$$\begin{aligned} \mathbb{D}_{\mathbf{u}} f(x, y) &= \nabla f(x, y) \cdot \mathbf{u} \\ &= |\nabla f| |\mathbf{u}| \cos \theta \\ &= |\nabla f| \cos \theta, \end{aligned}$$

where θ is the angle between ∇f and u. The maximum value of $\cos \theta$ is 1 and this occurs when $\theta = 0$. Therefore, the maximum value of $\mathbb{D}_u f(x,y)$ is $|\nabla f|$ and it occurs when $\theta = 0$ i.e., when u has the same direction as ∇f.

Similarly, the directional derivative of functions of three variables with unit $u = (u_1, u_2, u_3)$ can be written as

$$\mathbb{D}_{\mathbf{u}} f(x, y, z) = \nabla f(x, y, z) \cdot \mathbf{u}.$$

We now introduce the concept of differentiability for functions of several variable, let's first recall the definition of differentiability in one variable case.

Let D be an open subset $\mathbb{R}$. The function $f: f : D \to \mathbb{R}$ is said to be differentiable at $x_0 \in D$ if

$$\lim_{x \to x_0} \frac{f(x) - f(x_0)}{x - x_0}$$

exists. The value of this limit is called the derivative of f at x_0 and is denoted by $f'x_0$.

The above definition may be restated as follows: The function $f : D \to \mathbb{R}$ is differentiable at $x_0 \in D$ if there is a number $f'(x_0)$ such that

$$\lim_{x \to x_0} \frac{\left| f(x) - f(x_0) - f'(x_0)(x - x_0) \right|}{\left| x - x_0 \right|} = 0.$$

Any real number a_0 determines a linear transformation $L : \mathbb{R} \to \mathbb{R}$ defined by

$$Lx = a_0 x.$$

In particular, $f'x_0$ determines a linear transformation $L : \mathbb{R} \to \mathbb{R}$ given by $Lx = f'(x_0)x$. Therefore, with this linear transformation, we may rewrite above equation as

$$\lim_{x \to x_0} \frac{\left| f(x) - f(x_0) - L(x - x_0) \right|}{\left| x - x_0 \right|} = 0.$$

We now use earlier equation to define differentiability of a function $f : \mathbb{R}^n \to \mathbb{R}^m$.

DEFINITION 10. (Differentiability) Let $D \subset \mathbb{R}^n$ be an open subset and let $f : D \to \mathbb{R}^m$. We say that f is differentiable at $x_0 \in D$ if there is a linear transformation $L : \mathbb{R}^n \to \mathbb{R}^m$ such that

$$\lim_{\mathbf{x} \to \mathbf{x_0}} \frac{\| f(\mathbf{x}) - f(\mathbf{x_0}) - L(\mathbf{x} - \mathbf{x_0}) \|}{\| \mathbf{x} - \mathbf{x_0} \|} = 0.$$

The linear transformation L of above equ. is called the derivative of f at x_0. We denote it by $f'x_0$.

We say that f is differentiable in D if it is differentiable at each every point of D.

DEFINITION 11. (Jacobian matrix) Let $f : D \subset \mathbb{R}^n \to \mathbb{R}^m$ is defined by

$$f(\mathbf{x}) = (f_1(\mathbf{x}), \ldots, f_m(\mathbf{x})),$$

where $f_i : D \to \mathbb{R}, 1 \le i \le m$. For each $x \in D$, we define the Jacobian matrix of f at x by

$$J_f(\mathbf{x}) := (a_{ij}),$$

where $a_{ij} = (\partial f_i / \partial x_j)(x), 1 \le i \le m, 1 \le j \le n$.

EXAMPLE

Let $f : \mathbb{R}^2 \to \mathbb{R}^3$ be given by

$$f(x_1, x_2) = (x_1^2, x_1 x_2, x_2^2).$$

Here, $f_1(x_1, x_2) = x_1^2$, $f_2(x_1, x_2) = x_1 x_2$, $f_3(x_1, x_2) = x_2^2$. Then

$$\frac{\partial f_1}{\partial x_1} = 2x_1, \quad \frac{\partial f_2}{\partial x_1} = x_2, \quad \frac{\partial f_3}{\partial x_1} = 0.$$

$$\frac{\partial f_1}{\partial x_2} = 0, \quad \frac{\partial f_2}{\partial x_2} = x_1, \quad \frac{\partial f_3}{\partial x_2} = 2x_2$$

Therefore,

$$J_f(x_1, x_2) = \begin{bmatrix} 2x_1 & 0 \\ x_2 & x_1 \\ 0 & 2x_2 \end{bmatrix}.$$

The following theorem gives a formula for computing derivative.

THEOREM: (Computing derivative) Let D be an open subset of $\mathbb{R}^n$ and $f : D \to \mathbb{R}^m$ be defined by

$$f(\mathbf{x}) = (f_1(\mathbf{x}), \dots, f_m(\mathbf{x})),$$

where $f_i = D \to \mathbb{R}, 1 \le i \le m$. If f is differentiable at a point x_0 in D, then each of the partial derivatives $(\partial f_i / \partial x_j)(x_0)$ exists, $1 \le i \le m$, $1 \le j \le n$. Furthermore,

$$f'(\mathbf{x_0}) = J_f(\mathbf{x_0}).$$

Note that the linear transformation L is defined by the Jacobian matrix of f at x_0. In particular, for m = 1, we have

$$L = f'(\mathbf{x_0}) = \nabla f(\mathbf{x_0}).$$

The following theorem gives the suffcient condition for differentiability of f.

THEOREM: (Sufficient condition for differentiability) Let $D \subset \mathbb{R}^n$ be an open set and $f : D \to \mathbb{R}^m$ be defined by

$$f(\mathbf{x}) = (f_1(\mathbf{x}), \dots, f_m(\mathbf{x})),$$

where $f_i = D \to \mathbb{R}, 1 \le i \le m$. Suppose that $(\partial f_i / \partial x_j)(x_o)$ exists and continuous on D, $1 \le i \le m$, $1 \le j \le n$. Then f is differentiable on D.

We shall stating some results on maxima and minima in the case of a function of several variables. We restrict our discussion to functions of two variables only.

DEFINITION: (Maxima and Minima) Let $f(x, y)$ be a function of two variables. A point (x_0, y_0) is a local minimum point for f if there is a disk $D_\delta(x_0, y_0)$ about (x_0, y_0) such that

$$f(x, y) \ge f(x_0, y_0) \ \ for \ all \ (x, y) \in D_\delta(x_0, y_0).$$

The number $f(x_0, y_0)$ is called a local minimum value.

Similarly, if there is a disk $D_\delta(x_0, y_0)$ about (x_0, y_0) such that

$$f(x, y) \le f(x_0, y_0) \ \ for \ all \ (x, y) \in D_\delta(x_0, y_0)$$

then the point (x_0, y_0) a local maximum point for f.

A point which is either a local maximum or minimum point is called a local extremum.

The following is the analog in two variables of the first derivative test for one variable.

First Derivative Test

If (x_0, y_0) is a local extremum of f and the partial derivatives of f exist at (x_0, y_0), then

$$f_x(x_0, y_0) = f_y(x_0, y_0) = 0.$$

Such point (x_0, y_0) is also called a critical point of f.

Second Derivative Test

Let $f(x, y)$ have continuous second-order partial derivatives, and suppose that (x_0, y_0) is a critical point for f. Then

$$f_x(x_0, y_0) = 0 \ \ and \ \ f_y(x_0, y_0) = 0$$

Let $A = f_{xx}(x_0, y_0), B = f_{xy}(x_0, y_0), and \ C = f_{yy}(x_0, y_0)$. Then the following statements are true.

(a) If $A > 0, AC - B^2 > 0$ than (x_0, y_0) is a local minimum.
(b) If $A < 0, AC - B^2 > 0$ than (x_0, y_0) is a local maximum.
(c) If $AC - B^2 < 0$ than (x_0, y_0) is a saddle point
(d) If $AC - B^2 = 0$ than the test is inconclusive.

Initial Value Problem

In mathematics, in the field of differential equations, an initial value problem (also called the Cauchy problem by some authors) is an ordinary differential equation together with a specified value, called the initial condition, of the unknown function at a given point in the domain of the solution. In physics or other sciences, modeling a system frequently amounts to solving an initial value problem; in this context, the differential equation is an evolution equation specifying how, given initial conditions, the system will evolve with time.

Definition

An initial value problem is a differential equation

$$y'(t) = f(t, y(t)) \text{ with } f : \Omega \subset \mathbb{R} \times \mathbb{R}^n \to \mathbb{R}^n \text{ where } \Omega \text{ is an open set of } \mathbb{R} \times \mathbb{R}^n,$$

together with a point in the domain of f

$$(t_0, y_0) \in \Omega$$
,

called the initial condition.

A solution to an initial value problem is a function y that is a solution to the differential equation and satisfies.

$$y(t_0) = y_0.$$

In higher dimensions, the differential equation is replaced with a family of equations $y_i'(t) = f_i(t, y_1(t), y_2(t), ...)$, and $y(t)$ is viewed as the vector $(y_1(t), ..., y_n(t))$. More generally, the unknown function y can take values on infinite dimensional spaces, such as Banach spaces or spaces of distributions.

Initial value problems are extended to higher orders by treating the derivatives in the same way as an independent function, e.g. $y''(t) = f(t, y(t), y'(t))$.

Existence and Uniqueness of Solutions

For a large class of initial value problems, the existence and uniqueness of a solution can be illustrated through the use of a calculator.

The Picard–Lindelöf theorem guarantees a unique solution on some interval containing t_0 if f is continuous on a region containing t_0 and y_0 and satisfies the Lipschitz condition on the variable y. The proof of this theorem proceeds by reformulating the problem as an equivalent integral equation. The integral can be considered an operator which maps one function into another, such that the solution is a fixed point of the operator. The Banach fixed point theorem is then invoked to show that there exists a unique fixed point, which is the solution of the initial value problem.

An older proof of the Picard–Lindelöf theorem constructs a sequence of functions which converge to the solution of the integral equation, and thus, the solution of the initial value problem. Such a construction is sometimes called "Picard's method" or "the method of successive approximations". This version is essentially a special case of the Banach fixed point theorem.

Hiroshi Okamura obtained a necessary and sufficient condition for the solution of an initial value problem to be unique. This condition has to do with the existence of a Lyapunov function for the system.

In some situations, the function f is not of class C^1, or even Lipschitz, so the usual result guaranteeing the local existence of a unique solution does not apply. The Peano existence theorem however proves that even for f merely continuous, solutions are guaranteed to exist locally in time; the problem is that there is no guarantee of uniqueness. An even more general result is the Carathéodory existence theorem, which proves existence for some discontinuous functions f.

Examples

A simple example is to solve $y' = 0.85y$ and $y(0) = 19$. We are trying to find a formula for $y(t)$ that satisfies these two equations.

Start by noting that $y' = \dfrac{dy}{dt}$, so

$$\frac{dy}{dt} = 0.85y$$

Now rearrange the equation so that y is on the left and t on the right

$$\frac{dy}{y} = 0.85dt$$

Now integrate both sides (this introduces an unknown constant B).

$$\ln|y| = 0.85t + B$$

Eliminate the ln

$$|y| = e^{B}e^{0.85t}$$

Let C be a new unknown constant, $C = \pm e^{B}$, so

$$y = Ce^{0.85t}$$

Now we need to find a value for C. Use $y(0) = 19$ as given at the start and substitute 0 for t and 19 for y

$$19 = Ce^{0.85 \cdot 0}$$

$$C = 19$$

this gives the final solution of $y(t) = 19e^{0.85t}$.

Second Example

The solution of

$$y' + 3y = 6t + 5, \qquad y(0) = 3$$

can be found to be

$$y(t) = 2e^{-3t} + 2t + 1.$$

Indeed,

$$\begin{aligned} y' + 3y &= \tfrac{d}{dt}(2e^{-3t} + 2t + 1) + 3(2e^{-3t} + 2t + 1) \\ &= (-6e^{-3t} + 2) + (6e^{-3t} + 6t + 3) \\ &= 6t + 5. \end{aligned}$$

Initial Condition

In mathematics and particularly in dynamic systems, an initial condition, in some contexts called a seed value, is a value of an evolving variable at some point in time designated as the initial time (typically denoted $t = 0$). For a system of order k (the number of time lags in discrete time, or the order of the largest derivative in continuous time) and dimension n (that is, with n different evolving variables, which together can be denoted by an n-dimensional coordinate vector), generally nk initial conditions are needed in order to trace the system's variables forward through time.

In both differential equations in continuous time and difference equations in discrete time, initial conditions affect the value of the dynamic variables (state variables) at any future time. In continuous time, the problem of finding a closed form solution for the state variables as a function of time and of the initial conditions is called the initial value problem. A corresponding problem exists for discrete time situations. While a closed form solution is not always possible to obtain, future values of a discrete time system can be found by iterating forward one time period per iteration, though rounding error may make this impractical over long horizons.

Linear System

Discrete Time

A linear matrix difference equation of the homogeneous (having no constant term) form $X_{t+1} = AX_t$ has closed form solution $X_t = A^t X_0$ predicated on the vector X_0 of initial conditions on the individual variables that are stacked into the vector; X_0 is called the vector of initial conditions or simply the initial condition, and contains nk pieces of information, n being the dimension of the vector X and $k = 1$ being the number of time lags in the system. The initial conditions in this linear system do not affect the qualitative nature of the future behavior of the state variable X; that behavior is stable or unstable based on the eigenvalues of the matrix A but not based on the initial conditions.

Alternatively, a dynamic process in a single variable x having multiple time lags is

$$x_t = a_1 x_{t-1} + a_2 x_{t-2} + \cdots + a_k x_{t-k}.$$

Here the dimension is $n = 1$ and the order is k, so the necessary number of initial conditions to trace the system through time, either iteratively or via closed form solution, is $nk = k$. Again the initial conditions do not affect the qualitative nature of the variable's long-term evolution. The solution of this equation is found by using its characteristic equation $\lambda^k - a_1 \lambda^{k-1} - a_2 \lambda^{k-2} - \cdots - a_{k-1} \lambda - a_k = 0$ to obtain the latter's k solutions, which are the characteristic values $\lambda_1, \ldots, \lambda_k$, for use in the solution equation

$$x_t = c_1 \lambda_1^t + \cdots + c_k \lambda_k^t.$$

Here the constants $c_1, \ldots, c_k$ are found by solving a system of k different equations based on this equation, each using one of k different values of t for which the specific initial condition x_t is known.

Continuous Time

A differential equation system of the first order with n variables stacked in a vector X is

$$\frac{dX}{dt} = AX.$$

Its behavior through time can be traced with a closed form solution conditional on an initial condition vector X_0. The number of required initial pieces of information is the dimension n of the system times the order $k = 1$ of the system, or n. The initial conditions do not affect the qualitative behavior (stable or unstable) of the system.

A single k^{th} order linear equation in a single variable x is

$$\frac{d^k x}{dt^k} + a_{k-1}\frac{d^{k-1}x}{dt^{k-1}} + \cdots + a_1\frac{dx}{dt} + a_0 x = 0.$$

Here the number of initial conditions necessary for obtaining a closed form solution is the dimension $n = 1$ times the order k, or simply k. In this case the k initial pieces of information will typically not be different values of the variable x at different points in time, but rather the values of x and its first $k - 1$ derivatives, all at some point in time such as time zero. The initial conditions do not affect the qualitative nature of the system's behavior. The characteristic equation of this dynamic equation is $\lambda^k + a_{k-1}\lambda^{k-1} + \cdots + a_1\lambda + a_0 = 0$, whose solutions are the characteristic values $\lambda_1, \ldots, \lambda_k$; these are used in the solution equation

$$x(t) = c_1 e^{\lambda_1 t} + \cdots + c_k e^{\lambda_k t}.$$

This equation and its first $k - 1$derivatives form a system of k equations that can be solved for the k parameters $c_1, \ldots, c_k$, given the known initial conditions on x and its $k - 1$ derivatives' values at some time t.

Nonlinear Systems

Nonlinear systems can exhibit a substantially richer variety of behavior than can linear systems. In particular, the initial conditions can affect whether the system diverges to infinity or whether it converges to one or another attractor of the system. Each attractor, a (possibly disconnected) region of values that some dynamic paths approach but never leave, has a (possibly disconnected) basin of attraction such that state variables with initial conditions in that basin (and nowhere else) will evolve toward that attractor. Even nearby initial conditions could be in basins of attraction of different attractors.

Moreover, in those nonlinear systems showing chaotic behavior, the evolution of the variables exhibits sensitive dependence on initial conditions: the iterated values of any two very nearby points on the same strange attractor, while each remaining on the attractor, will diverge from each other over time. Thus even on a single attractor the precise values of the initial conditions make a substantial difference for the future positions of the iterates. This feature makes accurate simulation of future values difficult, and impossible over long horizons, because stating the initial conditions with exact precision is seldom possible and because rounding error is inevitable after even only a few iterations from an exact initial condition.

Lipschitz Continuity

In mathematical analysis, Lipschitz continuity, named after Rudolf Lipschitz, is a strong form of uniform continuity for functions. Intuitively, a Lipschitz continuous function is limited in how fast it can change: there exists a definite real number such that, for every pair of points on the graph of this function, the absolute value of the slope of the line connecting them is not greater than this real number; this bound is called a *Lipschitz constant* of the function (or *modulus of uniform continuity*). For instance, every function that has bounded first derivatives is Lipschitz.

In the theory of differential equations, Lipschitz continuity is the central condition of the Picard–Lindelöf theorem which guarantees the existence and uniqueness of the solution to an initial value problem. A special type of Lipschitz continuity, called contraction, is used in the Banach fixed point theorem.

We have the following chain of inclusions for functions over a closed and bounded subset of the real line

Continuously differentiable ⊆ Lipschitz continuous ⊆ α-Hölder continuous ⊆ uniformly continuous ⊆ continuous

where $0 < \alpha \leq 1$. We also have

Lipschitz continuous ⊆ absolutely continuous ⊆ bounded variation ⊆ differentiable almost everywhere

Definitions

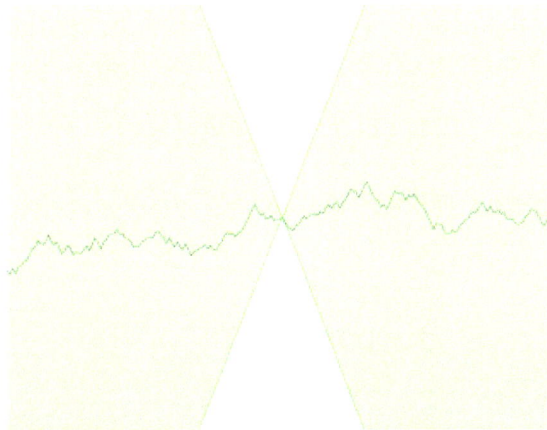

For a Lipschitz continuous function, there is a double cone (shown in white) whose vertex can be translated along the graph, so that the graph always remains entirely outside the cone.

Given two metric spaces (X, d_X) and (Y, d_Y), where d_X denotes the metric on the set X and d_Y is the metric on set Y, a function $f: X \to Y$ is called Lipschitz continuous if there exists a real constant $K \geq 0$ such that, for all x_1 and x_2 in X,

$$d_Y(f(x_1), f(x_2)) \leq K d_X(x_1, x_2).$$

Any such K is referred to as a Lipschitz constant for the function f. The smallest constant is sometimes called the (best) Lipschitz constant; however, in most cases, the latter notion is less relevant. If $K = 1$ the function is called a short map, and if $0 \leq K < 1$ the function is called a contraction.

In particular, a real-valued function $f: R \to R$ is called Lipschitz continuous if there exists a positive real constant K such that, for all real x_1 and x_2,

$$| f(x_1) - f(x_2) | \le K | x_1 - x_2 |.$$

In this case, Y is the set of real numbers R with the metric $d_Y(y_1, y_2) = |y_1 - y_2|$, and X might be a subset of R.

In general, the inequality is (trivially) satisfied if $x_1 = x_2$. Otherwise, one can equivalently define a function to be Lipschitz continuous if and only if there exists a constant $K \ge 0$ such that, for all $x_1 \ne x_2$,

$$\frac{d_Y(f(x_1), f(x_2))}{d_X(x_1, x_2)} \le K.$$

For real-valued functions of several real variables, this holds if and only if the absolute value of the slopes of all secant lines are bounded by K. The set of lines of slope K passing through a point on the graph of the function forms a circular cone, and a function is Lipschitz if and only if the graph of the function everywhere lies completely outside of this cone.

A function is called locally Lipschitz continuous if for every x in X there exists a neighborhood U of x such that f restricted to U is Lipschitz continuous. Equivalently, if X is a locally compact metric space, then f is locally Lipschitz if and only if it is Lipschitz continuous on every compact subset of X. In spaces that are not locally compact, this is a necessary but not a sufficient condition.

More generally, a function f defined on X is said to be Hölder continuous or to satisfy a Hölder condition of order $\alpha > 0$ on X if there exists a constant $M > 0$ such that

$$d_Y(f(x), f(y)) \le M d_X(x, y)^\alpha$$

for all x and y in X. Sometimes a Hölder condition of order α is also called a uniform Lipschitz condition of order $\alpha > 0$.

If there exists a $K \ge 1$ with

$$\frac{1}{K} d_X(x_1, x_2) \le d_Y(f(x_1), f(x_2)) \le K d_X(x_1, x_2)$$

then f is called bilipschitz (also written bi-Lipschitz). A bilipschitz mapping is injective, and is in fact a homeomorphism onto its image. A bilipschitz function is the same thing as an injective Lipschitz function whose inverse function is also Lipschitz.

Examples

Lipschitz continuous functions

- The function $f(x) = \sqrt{x2 + 5}$ defined for all real numbers is Lipschitz continuous with the Lipschitz constant $K = 1$, because it is everywhere differentiable and the absolute value of the derivative is bounded above by 1.

- Likewise, the sine function is Lipschitz continuous because its derivative, the cosine function, is bounded above by 1 in absolute value.

- The function $f(x) = |x|$ defined on the reals is Lipschitz continuous with the Lipschitz constant equal to 1, by the reverse triangle inequality. This is an example of a Lipschitz continuous function that is not differentiable. More generally, a norm on a vector space is Lipschitz continuous with respect to the associated metric, with the Lipschitz constant equal to 1.

Lipschitz continuous functions that are not everywhere differentiable

- The function $f(x) = |x|$.

Continuous functions that are not (globally) Lipschitz continuous

- The function $f(x) = \sqrt{x}$ defined on [0, 1] is *not* Lipschitz continuous. This function becomes infinitely steep as x approaches 0 since its derivative becomes infinite. However, it is uniformly continuous as well as Hölder continuous of class $C^{0,\alpha}$ for $\alpha \leq 1/2$.

Differentiable functions that are not (globally) Lipschitz continuous

- The function $f(x) = x^{3/2}\sin(1/x)$ where $x \neq 0$ and $f(0) = 0$, restricted on [0, 1], gives an example of a function that is differentiable on a compact set while not locally Lipschitz because its derivative function is not bounded.

Analytic functions that are not (globally) Lipschitz continuous

- The exponential function becomes arbitrarily steep as $x \to \infty$, and therefore is *not* globally Lipschitz continuous, despite being an analytic function.

- The function $f(x) = x^2$ with domain all real numbers is *not* Lipschitz continuous. This function becomes arbitrarily steep as x approaches infinity. It is however locally Lipschitz continuous.

Properties

- An everywhere differentiable function $g : R \to R$ is Lipschitz continuous (with $K = \sup |g'(x)|$) if and only if it has bounded first derivative; one direction follows from the mean value theorem. In particular, any continuously differentiable function is locally Lipschitz, as continuous functions are locally bounded so its gradient is locally bounded as well.

- A Lipschitz function $g : R \to R$ is absolutely continuous and therefore is differentiable almost everywhere, that is, differentiable at every point outside a set of Lebesgue measure zero. Its derivative is essentially bounded in magnitude by the Lipschitz constant, and for $a < b$, the difference $g(b) - g(a)$ is equal to the integral of the derivative g' on the interval $[a, b]$.

  o Conversely, if $f : I \to R$ is absolutely continuous and thus differentiable almost everywhere, and satisfies $|f'(x)| \leq K$ for almost all x in I, then f is Lipschitz continuous with Lipschitz constant at most K.

  o More generally, Rademacher's theorem extends the differentiability result to Lipschitz mappings between Euclidean spaces: a Lipschitz map $f : U \to R^m$, where U is an open set in R^n, is almost everywhere differentiable. Moreover, if K is the best Lipschitz constant of f, then $\| Df(x) \| \leq K$ whenever the total derivative Df exists.

- For a differentiable Lipschitz map $f : U \to \mathbf{R}^m$ the inequality $\|Df\|_{\infty, U} \leq K$ holds for the best Lipschitz constant of f, and it turns out to be an equality if the domain U is convex.

- Suppose that $\{f_n\}$ is a sequence of Lipschitz continuous mappings between two metric spaces, and that all f_n have Lipschitz constant bounded by some K. If f_n converges to a mapping f uniformly, then f is also Lipschitz, with Lipschitz constant bounded by the same K. In particular, this implies that the set of real-valued functions on a compact metric space with a particular bound for the Lipschitz constant is a closed and convex subset of the Banach space of continuous functions. This result does not hold for sequences in which the functions may have *unbounded* Lipschitz constants, however. In fact, the space of all Lipschitz functions on a compact metric space is a subalgebra of the Banach space of continuous functions, and thus dense in it, an elementary consequence of the Stone–Weierstrass theorem (or as a consequence of Weierstrass approximation theorem, because every polynomial is Lipschitz continuous).

- Every Lipschitz continuous map is uniformly continuous, and hence *a fortiori* continuous. More generally, a set of functions with bounded Lipschitz constant forms an equicontinuous set. The Arzelà–Ascoli theorem implies that if $\{f_n\}$ is a uniformly bounded sequence of functions with bounded Lipschitz constant, then it has a convergent subsequence. By the result of the previous paragraph, the limit function is also Lipschitz, with the same bound for the Lipschitz constant. In particular the set of all real-valued Lipschitz functions on a compact metric space X having Lipschitz constant $\leq K$ is a locally compact convex subset of the Banach space $C(X)$.

- For a family of Lipschitz continuous functions f_α with common constant, the function $\sup_\alpha f_\alpha$ (and $\inf_\alpha f_\alpha$) is Lipschitz continuous as well, with the same Lipschitz constant, provided it assumes a finite value at least at a point.

- If U is a subset of the metric space M and $f : U \to \mathbf{R}$ is a Lipschitz continuous function, there always exist Lipschitz continuous maps $M \to \mathbf{R}$ which extend f and have the same Lipschitz constant as f. An extension is provided by

$$\tilde{f}(x) := \inf_{u \in U}\{f(u) + k\,d(x,u)\},$$

where k is a Lipschitz constant for f on U.

Lipschitz Manifolds

Let U and V be two open sets in $\mathbf{R}^n$. A function $T : U \to V$ is called bi-Lipschitz if it is a Lipschitz homeomorphism onto its image, and its inverse is also Lipschitz.

Using bi-Lipschitz mappings, it is possible to define a Lipschitz structure on a topological manifold, since there is a pseudogroup structure on bi-Lipschitz homeomorphisms. This structure is intermediate between that of a piecewise-linear manifold and a smooth manifold. In fact a PL structure gives rise to a unique Lipschitz structure; it can in that sense 'nearly' be smoothed.

One-sided Lipschitz

Let F(x) be an upper semi-continuous function of x, and that F(x) is a closed, convex set for all x. Then F is one-sided Lipschitz if

$$(x_1 - x_2)^T (F(x_1) - F(x_2)) \leq C \| x_1 - x_2 \|^2$$

for some C for all x_1 and x_2.

It is possible that the function F could have a very large Lipschitz constant but a moderately sized, or even negative, one-sided Lipschitz constant. For example, the function

$$\begin{cases} F : \mathbf{R}^2 \to \mathbf{R}, \\ F(x, y) = -50(y - \cos(x)) \end{cases}$$

has Lipschitz constant $K = 50$ and a one-sided Lipschitz constant $C = 0$. An example which is one-sided Lipschitz but not Lipschitz continuous is $F(x) = e^{-x}$, with $C = 0$.

Existence and Uniqueness of Solutions

There are many instances where a physical problem is represented by differential equations may be with initial or boundary conditions. The existence of solutions for mathematical models is vital as otherwise it may not be relevant to the physical problem. This tells us that existence of solutions is a fundamental problem. The chapter describes a few methods for establishing the existence of solutions, naturally under certain premises. We first look into a few preliminaries for the ensuing discussions. Let us now consider a class of functions satisfying Lipschitz condition, which plays an important role in the qualitative theory of differential equations. Its applications in showing the existence of a unique solution and continuous dependence on initial conditions are dealt with in this chapter.

Definition 1.1.1. A real valued function $f : D \to \mathbb{R}$ defined in a region $D \subset \mathbb{R}^2$ is said to satisfy Lipschitz condition in the variable x with a Lipschitz constant K, if the in equality

$$|f(t, x_1) - f(t, x_2)| \leq K |x_1 - x_2|, \tag{1}$$

holds whenever $(t, x_1), (t, x_2)$ are in D. In such a case, we say that f is a member of the class Lip(D, K).

As a consequence of Definition 1.1.1, a function f satisfies Lipschitz condition if and only if there exists a constant K > 0 such that

$$\frac{|f(t, x_1) - f(t, x_2)|}{|x_1 - x_2|} \leq K, \quad x_1 \neq x_2,$$

whenever $(t, x_1), (t, x_2)$ belong to D. Now we wish to find a general criteria which would ensure the Lipschitz condition on f. The following is a result in this direction. For simplicity, we assume the region D to be a closed rectangle.

Theorem. Define a rectangle R by

$$R = \{(t, x) : |t - t_0| \leq p, \ |x - x_0| \leq q\},$$

where p, q are some positive real numbers. Let $f : R \to \mathbb{R}$ be a real valued continues function. Let $\frac{\partial f}{\partial x}$ be defined and continuous on R. Then, f satisfies the Lipschitz condition on R.

Proof. Since $\frac{\partial f}{\partial x}$ is continuous on R, we have a positive constant A such that

$$\left| \frac{\partial f}{\partial x}(t, x) \right| \leq A, \tag{2}$$

for all $(t, x) \in R$. Let (t, x_1), (t, x_2) be any two points in R. By the mean value theorem of differential calculus, there exists a number s which lies between x_1 and x_2 such that

$$f(t, x_1) - f(t, x_2) = \frac{\partial f}{\partial x}(t, s)(x_1 - x_2).$$

Since the point $(t, x) \in R$ and by the inequality (2), we have

$$\left| \frac{\partial f}{\partial x}(t, s) \right| \leq A,$$

or else, we have

$$|f(t, x_1) - f(t, x_2)| \leq A|x_1 - x_2|,$$

whenever (t, x_1), (t, x_2) are in R. The proof is complete.

The following example illustrates that the existence of partial derivative of f is not necessary for f to be a Lipschitz function.

Example. Let $R = \{(t, x) : |t| \leq 1, \ |x| \leq 1\}$ and let

$$f(t, x) = |x| \ \text{for} \ (t, x) \in R.$$

Then, the partial derivative of f at (t, 0) fails to exist but f satisfies Lipschitz condition in x on R with Lipschitz constant K = 1.

The example below shows that there are functions which do not satisfy the Lipschitz condition.

Example. Let f be defined by

$$f(t, x) = x^{1/2}$$

on the rectangle R = $\{(t, x) : |t| \leq 2, \ |x| \leq 2\}$. Then, f does not satisfy the inequality (1) in R. This is because

$$\frac{f(t, x) - f(t, 0)}{x - 0} = x^{-1/2}, \quad x \neq 0,$$

is unbounded in R.

If we alter the domain in the Example 1.1.4, f may satisfy the Lipschitz condition. e.g., take R = {(t, x) : |t| ≤ 2, 2 ≤ |x| ≤ 4} in Example 1.1.4.

Grönwall's Inequality

In mathematics, Grönwall's inequality (also called Grönwall's lemma or the Grönwall–Bellman inequality) allows one to bound a function that is known to satisfy a certain differential or integral inequality by the solution of the corresponding differential or integral equation. There are two forms of the lemma, a differential form and an integral form. For the latter there are several variants.

Grönwall's inequality is an important tool to obtain various estimates in the theory of ordinary and stochastic differential equations. In particular, it provides a comparison theorem that can be used to prove uniqueness of a solution to the initial value problem.

It is named for Thomas Hakon Grönwall (1877–1932). Grönwall is the Swedish spelling of his name, but he spelled his name as Gronwall in his scientific publications after emigrating to the United States.

The differential form was proven by Grönwall in 1919. The integral form was proven by Richard Bellman in 1943.

A nonlinear generalization of the Grönwall–Bellman inequality is known as Bihari–LaSalle inequality. Other variants and generalizations can be found in Pachpatte, B.G. (1998).

Differential Form

Let I denote an interval of the real line of the form $[a, \infty)$ or $[a, b]$ or $[a, b)$ with $a < b$. Let β and u be real-valued continuous functions defined on I. If u is differentiable in the interior I° of I (the interval I without the end points a and possibly b) and satisfies the differential inequality

$$u'(t) \leq \beta(t)u(t), \qquad t \in I^\circ,$$

then u is bounded by the solution of the corresponding differential *equation* $y'(t) = \beta(t)y(t)$:

$$u(t) \leq u(a)\exp\left(\int_a^t \beta(s)ds\right)$$

for all $t \in I$.

Remark: There are no assumptions on the signs of the functions β and u.

Proof

Define the function

$$v(t) = \exp\left(\int_a^t \beta(s)ds\right), \qquad t \in I.$$

Note that v satisfies

$$v'(t) = \beta(t)v(t), \qquad t \in I^{\circ},$$

with $v(a) = 1$ and $v(t) > 0$ for all $t \in I$. By the quotient rule

$$\frac{d}{dt}\frac{u(t)}{v(t)} = \frac{u'(t)v(t) - v'(t)u(t)}{v^2(t)} = \frac{u'(t)v(t) - \beta(t)v(t)u(t)}{v^2(t)} \leq 0, \qquad t \in I^{\circ},$$

Thus the derivative of the function $u(t)/v(t)$ is non-positive and the function is bounded above by its value at the initial point a of the interval I:

$$\frac{u(t)}{v(t)} \leq \frac{u(a)}{v(a)} = u(a), \qquad t \in I,$$

which is Grönwall's inequality.

Integral form for Continuous Functions

Let I denote an interval of the real line of the form $[a, \infty)$ or $[a, b]$ or $[a, b)$ with $a < b$. Let α, β and u be real-valued functions defined on I. Assume that β and u are continuous and that the negative part of α is integrable on every closed and bounded subinterval of I.

- (a) If β is non-negative and if u satisfies the integral inequality

$$u(t) \leq \alpha(t) + \int_a^t \beta(s)u(s)ds, \qquad \forall t \in I,$$

  then

$$u(t) \leq \alpha(t) + \int_a^t \alpha(s)\beta(s)\exp\left(\int_s^t \beta(r)dr\right)ds, \qquad t \in I.$$

- (b) If, in addition, the function α is non-decreasing, then

$$u(t) \leq \alpha(t)\exp\left(\int_a^t \beta(s)ds\right), \qquad t \in I.$$

Remarks

- There are no assumptions on the signs of the functions α and u.
- Compared to the differential form, differentiability of u is not needed for the integral form.
- For a version of Grönwall's inequality which doesn't need continuity of β and u.

Proof

(a) Define

$$v(s) = \exp\left(-\int_a^s \beta(r)dr\right)\int_a^s \beta(r)u(r)dr, \qquad s \in I.$$

Using the product rule, the chain rule, the derivative of the exponential function and the fundamental theorem of calculus, we obtain for the derivative

$$v'(s) = \underbrace{\left(u(s) - \int_a^s \beta(r)u(r)dr\right)}_{\leq \alpha(s)}\beta(s)\exp\left(-\int_a^s \beta(r)dr\right), \qquad s \in I,$$

where we used the assumed integral inequality for the upper estimate. Since β and the exponential are non-negative, this gives an upper estimate for the derivative of v. Since $v(a) = 0$, integration of this inequality from a to t gives

$$v(t) \leq \int_a^t \alpha(s)\beta(s)\exp\left(-\int_a^s \beta(r)dr\right)ds.$$

Using the definition of $v(t)$ for the first step, and then this inequality and the functional equation of the exponential function, we obtain

$$\int_a^t \beta(s)u(s)ds = \exp\left(\int_a^t \beta(r)dr\right)v(t)$$

$$\leq \int_a^t \alpha(s)\beta(s)\exp\left(\underbrace{\int_a^t \beta(r)dr - \int_a^s \beta(r)dr}_{=\int_s^t \beta(r)dr}\right)ds.$$

Substituting this result into the assumed integral inequality gives Grönwall's inequality.

(b) If the function α is non-decreasing, then part (a), the fact $\alpha(s) \leq \alpha(t)$, and the fundamental theorem of calculus imply that

$$u(t) \leq \alpha(t) + \left(-\alpha(t)\exp\left(\int_s^t \beta(r)dr\right)\right)\Big|_{s=a}^{s=t}$$

$$= \alpha(t)\exp\left(\int_a^t \beta(r)dr\right), \qquad t \in I.$$

Integral form with Locally Finite Measures

Let I denote an interval of the real line of the form $[a, \infty)$ or $[a, b]$ or $[a, b)$ with $a < b$. Let α and u be measurable functions defined on I and let μ be a non-negative measure on the Borel σ-algebra of I satisfying $\mu([a, t]) < \infty$ for all $t \in I$ (this is certainly satisfied when μ is a locally finite measure). Assume that u is integrable with respect to μ in the sense that

$$\int_{[a,t)} |u(s)|\mu(ds) < \infty, \qquad t \in I,$$

and that u satisfies the integral inequality

$$u(t) \leq \alpha(t) + \int_{[a,t)} u(s)\mu(ds), \qquad t \in I.$$

If, in addition,

- the function α is non-negative or

- the function $t \mapsto \mu([a, t])$ is continuous for $t \in I$ and the function α is integrable with respect to μ in the sense that

$$\int_{[a,t)} |\alpha(s)| \mu(ds) < \infty, \qquad t \in I,$$

then u satisfies Grönwall's inequality

$$u(t) \leq \alpha(t) + \int_{[a,t)} \alpha(s) \exp\left(\mu(I_{s,t})\right) \mu(ds)$$

for all $t \in I$, where $I_{s,t}$ denotes to open interval (s, t).

Remarks

- There are no continuity assumptions on the functions α and u.

- The integral in Grönwall's inequality is allowed to give the value infinity.

- If α is the zero function and u is non-negative, then Grönwall's inequality implies that u is the zero function.

- The integrability of u with respect to μ is essential for the result. For a counterexample, let μ denote Lebesgue measure on the unit interval $[0, 1]$, define $u(0) = 0$ and $u(t) = 1/t$ for $t \in (0, 1]$, and let α be the zero function.

- Makes the stronger assumptions that α is a non-negative constant and u is bounded on bounded intervals, but doesn't assume that the measure μ is locally finite. Compared to the one given below, their proof does not discuss the behaviour of the remainder $R_n(t)$.

Special Cases

- If the measure μ has a density β with respect to Lebesgue measure, then Grönwall's inequality can be rewritten as

$$u(t) \leq \alpha(t) + \int_a^t \alpha(s)\beta(s) \exp\left(\int_s^t \beta(r)dr\right) ds, \qquad t \in I.$$

- If the function α is non-negative and the density β of μ is bounded by a constant c, then

$$u(t) \leq \alpha(t) + c\int_a^t \alpha(s) \exp\left(c(t-s)\right) ds, \qquad t \in I.$$

- If, in addition, the non-negative function α is non-decreasing, then

$$u(t) \leq \alpha(t) + c\alpha(t)\int_a^t \exp\left(c(t-s)\right) ds = \alpha(t)\exp(c(t-a)), \qquad t \in I.$$

Outline of Proof

The proof is divided into three steps. In idea is to substitute the assumed integral inequality into itself n times. This is done in Claim 1 using mathematical induction. In Claim 2 we rewrite the measure of a simplex in a convenient form, using the permutation invariance of product measures. In the third step we pass to the limit n to infinity to derive the desired variant of Grönwall's inequality.

Detailed Proof

Claim 1: Iterating the Inequality

For every natural number n including zero,

$$u(t) \le \alpha(t) + \int_{[a,t)} \alpha(s) \sum_{k=0}^{n-1} \mu^{\otimes k}(A_k(s,t)) \mu(ds) + R_n(t)$$

with remainder

$$R_n(t) := \int_{[a,t)} u(s) \mu^{\otimes n}(A_n(s,t)) \mu(ds), \qquad t \in I,$$

where

$$A_n(s,t) = \{(s_1,\ldots,s_n) \in I_{s,t}^n \mid s_1 < s_2 < \cdots < s_n\}, \qquad n \ge 1,$$

is an n-dimensional simplex and

$$\mu^{\otimes 0}(A_0(s,t)) := 1.$$

Proof of Claim 1

We use mathematical induction. For $n = 0$ this is just the assumed integral inequality, because the empty sum is defined as zero.

Induction step from n to $n + 1$: Inserting the assumed integral inequality for the function u into the remainder gives

$$R_n(t) \le \int_{[a,t)} \alpha(s) \mu^{\otimes n}(A_n(s,t)) \mu(ds) + \tilde{R}_n(t)$$

with

$$\tilde{R}_n(t) := \int_{[a,t)} \left(\int_{[a,q)} u(s) \mu(ds) \right) \mu^{\otimes n}(A_n(q,t)) \mu(dq), \qquad t \in I.$$

Using the Fubini–Tonelli theorem to interchange the two integrals, we obtain

$$\tilde{R}_n(t) = \int_{[a,t)} u(s) \underbrace{\int_{(s,t)} \mu^{\otimes n}(A_n(q,t)) \mu(dq)}_{= \mu^{\otimes n+1}(A_{n+1}(s,t))} \mu(ds) = R_{n+1}(t), \qquad t \in I.$$

Hence Claim 1 is proved for $n + 1$.

Claim 2: Measure of the Simplex

For every natural number n including zero and all $s < t$ in I

$$\mu^{\otimes n}(A_n(s,t)) \leq \frac{(\mu(I_{s,t}))^n}{n!}$$

with equality in case $t \mapsto \mu([a, t])$ is continuous for $t \in I$.

Proof of Claim 2

For $n = 0$, the claim is true by our definitions. Therefore, consider $n \geq 1$ in the following.

Let S_n denote the set of all permutations of the indices in $\{1, 2, \ldots, n\}$. For every permutation $\sigma \in S_n$ define

$$A_{n,\sigma}(s,t) = \{(s_1,\ldots,s_n) \in I_{s,t}^n \mid s_{\sigma(1)} < s_{\sigma(2)} < \cdots < s_{\sigma(n)}\}.$$

These sets are disjoint for different permutations and

$$\bigcup_{\sigma \in S_n} A_{n,\sigma}(s,t) \subset I_{s,t}^n.$$

Therefore,

$$\sum_{\sigma \in S_n} \mu^{\otimes n}(A_{n,\sigma}(s,t)) \leq \mu^{\otimes n}(I_{s,t}^n) = (\mu(I_{s,t}))^n.$$

Since they all have the same measure with respect to the n-fold product of μ, and since there are $n!$ permutations in S_n, the claimed inequality follows.

Assume now that $t \mapsto \mu([a, t])$ is continuous for $t \in I$. Then, for different indices $i, j \in \{1, 2, \ldots, n\}$, the set

$$\{(s_1,\ldots,s_n) \in I_{s,t}^n \mid s_i = s_j\}$$

is contained in a hyperplane, hence by an application of Fubini's theorem its measure with respect to the n-fold product of μ is zero. Since

$$I_{s,t}^n \subset \bigcup_{\sigma \in S_n} A_{n,\sigma}(s,t) \cup \bigcup_{1 \leq i < j \leq n} \{(s_1,\ldots,s_n) \in I_{s,t}^n \mid s_i = s_j\},$$

the claimed equality follows.

Proof of Grönwall's Inequality

For every natural number n, Claim 2 implies for the remainder of Claim 1 that

$$|R_n(t)| \leq \frac{(\mu(I_{a,t}))^n}{n!} \int_{[a,t)} |u(s)|\mu(ds), \qquad t \in I.$$

By assumption we have $\mu(I_{a,t}) < \infty$. Hence, the integrability assumption on u implies that

$$\lim_{n\to\infty} R_n(t) = 0, \qquad t \in I.$$

Claim 2 and the series representation of the exponential function imply the estimate

$$\sum_{k=0}^{n-1} \mu^{\otimes k}(A_k(s,t)) \le \sum_{k=0}^{n-1} \frac{\left(\mu(I_{s,t})\right)^k}{k!} \le \exp\left(\mu(I_{s,t})\right)$$

for all $s < t$ in I. If the function a is non-negative, then it suffices to insert these results into Claim 1 to derive the above variant of Grönwall's inequality for the function u.

In case $t \mapsto \mu([a, t])$ is continuous for $t \in I$, Claim 2 gives

$$\sum_{k=0}^{n-1} \mu^{\otimes k}(A_k(s,t)) = \sum_{k=0}^{n-1} \frac{\left(\mu(I_{s,t})\right)^k}{k!} \to \exp\left(\mu(I_{s,t})\right) \qquad \text{as } n \to \infty$$

and the integrability of the function a permits to use the dominated convergence theorem to derive Grönwall's inequality.

The integral inequality, due to Gronwall, plays a useful part in the study of several properties of ordinary differential equations. In particular, we propose to employ it to establish the uniqueness of solutions.

Theorem. (Gronwall inequality) Assume that $f, g : [t_0, \infty] \to \mathbb{R}_+$ are non-negative continuous functions. Let $k > 0$ be a constant. Then, the inequality

$$f(t) \le k + \int_{t_0}^t g(s)f(s)ds, \quad t \ge t_0,$$

implies the inequality

$$f(t) \le k \exp\left(\int_{t_0}^t g(s)ds\right), \quad t \ge t_0.$$

Proof. By hypotheses, we have

$$\frac{f(t)g(t)}{k + \int_{t_0}^t g(s)f(s)ds} \le g(t), \quad t \ge t_0.$$

(3)

Since,

$$\frac{d}{dt}\left(k + \int_{t_0}^t g(s)f(s)ds\right) = f(t)g(t),$$

by integrating (3) between the limits t_0 and t, we have

$$\ln\left(k + \int_{t_0}^{t} g(s)f(s)ds\right) - \ln k \leq \int_{t_0}^{t} g(s)ds.$$

In other words,

$$k + \int_{t_0}^{t} g(s)f(s)ds \leq k \exp\left(\int_{t_0}^{t} g(s)ds\right).$$

(4)

The inequality (4) together with the hypotheses leads to the desired conclusion.

An interesting and useful consequence is :

Corollary. Let f and k be as in Theorem. If the inequality

$$f(t) \leq k \int_{t_0}^{t} f(s)ds, \quad t \geq t_0,$$

holds then,

$$f(t) \equiv 0, \ for \ t \geq t_0.$$

Proof. For any $\epsilon > 0$, we have

$$f(t) < \epsilon + k \int_{t_0}^{t} f(s)ds, \quad t \geq t_0.$$

By Theorem, we have

$$f(t) < \epsilon \exp k(t - t_0), \quad t \geq t_0.$$

Since ϵ is arbitrary, we have f (t) ≡ 0 for t ≥ t$_0$.

Picard's Successive Approximations

In this section we define the Picard's Successive Approximations which is used later for showing the existence of a unique solution of an IVP under certain assumptions. Let $D \subset \mathbb{R}^2$ is an open connected set and $f : D \to \mathbb{R}$ is continuous in (t, x) on D. We begin with an initial value problem

$$x' = f(t, x), \quad x(t_0) = x_0.$$

(5)

Also let (t$_0$, x$_0$) be in D. Geometrically speaking, solving (5) is to find a function x whose graph passes through (t$_0$, x$_0$) and the slope of x coincides with f (t, x) whenever (t, x) belongs to some neighborhood of (t$_0$, x$_0$). Such a class of problems is called a local existence problem for an initial value problem. Unfortunately, the usual elementary procedures for determining solutions may not materialize for (5). The need perhaps is a sequential approach to costruct a solution x of (5). This is where the method of successive approximations finds its utility. The iterative procedure for solving

(5) is important and needs a bit of knowledge of real analysis. The key to the general theory is an equivalent representation of (5) by the 'integral equation'

$$x(t) = x_0 + \int_{t_0}^{t} f(s, x(s))ds. \tag{6}$$

Equation (6) is called an integral equation since the unknown function x also occurs under the integral sign. The ensuing result establishes the equivalence of (5) and (6).

Lemma. Let $I \subset \mathbb{R}$ be an interval. A function $x : I \to \mathbb{R}$ is a solution of (5) on I if and only if x is a solution of (6) on I.

Proof. If x is a solution of (5) then, it is easy to show that x satisfies (6). Let x be a solution of (6). Obviously $x(t_0) = x_0$. Differentiating both sides of (6), and noting that f is continuous in (t, x), we have

$$x'(t) = f(t, x(t)), \ t \in I,$$

which completes the proof.

We do recall that f is a continuous function on D. Now we are set to define approximations to a solution of (5). First of all we start with an approximation to a solution and improve it by iteration. It is expected that the sequence of iterations converge to a solution of (5) in the limit. The importance of equation (6) now springs up. In this connection, we exploit the fact that the estimates can be easily handled with integrals rather than with derivatives.

A rough approximation to a solution of (5) is just the constant function

$$x_0(t) \equiv x_0.$$

We may get a better approximation by substituting $x_0(t)$ on the right hand side of (6), thus obtaining a new approximation x_1 given by

$$x_1(t) = x_0 + \int_{t_0}^{t} f(s, x_0(s))ds,$$

as long as $(s, x_0(s)) \in D$. To get a still better approximation, we repeat the process thereby defining

$$x_2(t) = x_0 + \int_{t_0}^{t} f(s, x_1(s))ds,$$

as long as $(s, x_1(s)) \in D$. In general, we define x_n inductively by

$$x_n(t) = x_0 + \int_{t_0}^{t} f(s, x_{n-1}(s))ds, \quad n = 1, 2, \ldots, \tag{7}$$

as long as $(s, _{n-1}(s)) \in D$, x_n is known as the n-th successive approximation. In the literature this procedure is known as "Picard's method of successive approximations". In the sequence $\{x_n\}$ does converge to a unique solution of (5) provided f satisfies the Lipschitz condition. Befpre we conclude this section let us have a few examples.

Example 1. For the illustration of the method of successive approximations consider an IVP

$$x' = -x, \ x(0) = 1, \ t \geq 0.$$

It is equivalent to the integral equation

$$x(t) = 1 - \int_0^t x(s)ds.$$

Let us note $t_0 = 0$ and $x_0 = 1$. The zero-th approximation is given by $x_0(t) \equiv 1$. The first approximation is

$$x_1(t) = 1 - \int_0^t x_0(s)ds = 1 - t.$$

By the definition of the successive approximations, it follows that

$$x_2(t) = 1 - \left[\int_0^t (1 - s)ds \right] = 1 - \left[t - \frac{t^2}{2} \right].$$

In general, the n-th approximation is (use induction)

$$x_n(t) = 1 - t + \frac{t^2}{2} + \cdots + (-1)^n \frac{t^n}{n!}.$$

Let us note that x_n is the n-th partial sum of the power series for e^{-t}. It is easy to directly verify that e^{-t} is the solution of the IVP.

Example 2. Consider the IVP

$$x' = \frac{2x}{t}, \ t > 0, \ x'(0) = 0, \ x(0) = 0.$$

The zero-th approximation x_0 is identically zero because $x(0) = 0$. The first approximation is $x_1 \equiv 0$. Also we have

$$x_n \equiv 0, \quad \text{for all } n.$$

Thus, the sequence of functions $\{x_n\}$ converges to the identically zero function. Clearly $x \equiv 0$ is a solution of the IVP. On the other hand, it is not hard to check that

$$x(t) = t^2$$

is also a solution of the IVP which shows that if at all the successive approximations con- verges, they converge to one of the solutions of the IVP.

Picard Theorem

In complex analysis, Picard's great theorem and Picard's little theorem are related theorems about the range of an analytic function. They are named after Émile Picard.

The Theorems

Plot of the function $\exp(\frac{1}{z})$, centered on the essential singularity at $z = 0$. The hue of a point z represents the argument of $\exp(\frac{1}{z})$, the luminance represents its absolute value. This plot shows that arbitrarily close to the singularity, all non-zero values are attained.

Little Picard Theorem: If a function $f : C \to C$ is entire and non-constant, then the set of values that $f(z)$ assumes is either the whole complex plane or the plane minus a single point.

Sketch of Proof: Picard's original proof was based on properties of the modular lambda function, usually denoted by λ, and which performs, using modern terminology, the holomorphic universal covering of the twice punctured plane by the unit disc. This function is explicitly constructed in the theory of elliptic functions. If f omits two values, then the composition of f with the inverse of the modular function maps the plane into the unit disc which implies that f is constant by Liouville's theorem.

This theorem is a significant strengthening of Liouville's theorem which states that the image of an entire non-constant function must be unbounded. Many different proofs of Picard's theorem were later found and Schottky's theorem is a quantitative version of it. In the case where the values of f are missing a single point, this point is called a lacunary value of the function.

Great Picard's Theorem: If an analytic function f has an essential singularity at a point w, then on any punctured neighborhood of w, $f(z)$ takes on all possible complex values, with at most a single exception, infinitely often.

This is a substantial strengthening of the Casorati-Weierstrass theorem, which only guarantees that the range of f is dense in the complex plane. A result of the Great Picard Theorem is that any entire, non-polynomial function attains all possible complex values infinitely often, with at most one exception.

The "single exception" is needed in both theorems, as demonstrated here:

- e^z is an entire non-constant function that is never 0,

- $e^{1/z}$ has an essential singularity at 0, but still never attains 0 as a value.

Generalization and Current Research

Great Picard's theorem is true in a slightly more general form that also applies to meromorphic functions:

Great Picard's Theorem (meromorphic version): If M is a Riemann surface, w a point on M, $P^1(C)$ = $C \cup \{\infty\}$ denotes the Riemann sphere and $f: M\backslash\{w\} \rightarrow P^1(C)$ is a holomorphic function with essential singularity at w, then on any open subset of M containing w, the function $f(z)$ attains all but at most *two* points of $P^1(C)$ infinitely often.

Example: The meromorphic function $f(z) = 1/(1 - e^{1/z})$ has an essential singularity at $z = 0$ and attains the value ∞ infinitely often in any neighborhood of 0; however it does not attain the values 0 or 1.

With this generalization, *Little Picard Theorem* follows from *Great Picard Theorem* because an entire function is either a polynomial or it has an essential singularity at infinity. As with the little theorem, the (at most two) points that are not attained are lacunary values of the function.

The following conjecture is related to "Great Picard's Theorem":

Conjecture: Let $\{U_1, ..., U_n\}$ be a collection of open connected subsets of C that cover the punctured unit disk D \ {0}. Suppose that on each U_j there is an injective holomorphic function f_j, such that $df_j = df_k$ on each intersection $U_j \cap U_k$. Then the differentials glue together to a meromorphic 1-form on D.

It is clear that the differentials glue together to a holomorphic 1-form g dz on D \ {0}. In the special case where the residue of g at 0 is zero the conjecture follows from the "Great Picards's Theorem".

With all the remarks and examples, the reader may have a number of doubts about the effectiveness and utility of Picard's method in practice. It may be speculated whether the successive integrations are defined at all or whether they lead to complicated computations. However, we mention that Picard's method has made a landmark in the theory of differential equations. It gives not only a method to determine an approximate solution subject to a given error but also establishes the existence of a unique solution of initial value problems under general conditions.

In all of what follows we assume that the function $f: R \rightarrow \mathbb{R}$ is bounded by L and satisfies the Lipschitz condition with the Lipschitz constant K on the closed rectangle

$$R = \{(t, x) \in \mathbb{R}^2 : |t - t_0| \leq a, \ |x - x_0| \leq b, \ a > 0, b > 0\}.$$

Before proceeding further, we need to show that the successive approximations defined by (7) are well defined on an interval I. That is, to define x_{j+1} on I, it is necessary to show that $(s, x_j(s))$ lies in R, for each s in I and $j \geq 1$.

Lemma. Let $h = \min\left(a, \dfrac{b}{L}\right)$. Then, the successive approximations given by (7) are defined on I = $|t - t_0| \leq h$. Further,

$$|x_j(t) - x_0| \leq L|t - t_0| \leq b, \quad j = 1, 2, \ldots, t \in I. \tag{8}$$

Proof. The method of induction is used to prove the lemma. Since $(t_0, x_0) \in R$, obviously $x_0(t) \equiv x_0$ satisfies (8). By the induction hypothesis, let us assume that, for any $0 < j \leq n$, x_n is defined on I and satisfies (8). Consequently $(s, x_n(s)) \in R$, for all s in I. So, x_{n+1} is defined on I. By definition, we have

$$x_{n+1}(t) = x_0 + \int_{t_0}^{t} f(s, x_n(s))ds, \quad t \in I.$$

Using the induction hypothesis, it now follows that

$$|x_{n+1}(t) - x_0| = \left| \int_{t_0}^{t} f(s, x_n(s))ds \right| \leq \int_{t_0}^{t} |f(s, x_n(s))|ds \leq L|t - t_0| \leq Lh \leq b.$$

Thus, x_{n+1} satisfies (8). This completes the proof

We now state and prove the Picard's theorem, a fundamental result dealing with the problem of existence of a unique solution for a class of initial value problems ,as given by (5). Recall that the closed rectangle is defined in Lemma.

Theorem. (Picard's Theorem) Let $f : R \to \mathbb{R}$ be continuous and be bounded by L and satisfy Lipschitz condition with Lipschitz constant K on the closed rectangle R. Then, the successive approximations n = 1, 2, . . . , given by (7) converge uniformly on an interval

$$I : |t - t_0| \leq h, \ h = \min\left(a, \frac{b}{L}\right),$$

to a solution x of the IVP (15). In addition, this solution is unique.

Proof. We know that the IVP (5) is equivalent to the integral equation (6) and it is sufficient to show that the successive approximations x_n converge to a unique solution of (6) and hence, to the unique solution of the IVP (5). First, note that

$$x_n(t) = x_0(t) + \sum_{i=1}^{n} \left[x_i(t) - x_{i-1}(t)\right]$$

is the n-th partial sum of the series

$$x_0(t) + \sum_{i=1}^{\infty} \left[x_i(t) - x_{i-1}(t)\right] \tag{9}$$

The convergence of the sequence $\{x_n\}$ is equivalent to the convergence of the series (9). We complete the proof by showing that:

(a) the series (9) converges uniformly to a continuous function x(t);

(b) x satisfies the integral equation (6);

(c) x is the unique solution of (5).

To start with we fix a positive number $h = \min\left(a, \dfrac{b}{L}\right)$. By Lemma 1.2.1 the successive approximations x_n, n = 1, 2, ..., in (7) are well defined on I : $|t - t_0| \le h$. Henceforth, we stick to the interval $I^+ = [t_0, t_0 + h]$. The proof on the interval $I^- = [t_0 - h, t_0]$ is similar except for minor modifications.

We estimate $x_{j+1} - x_j$ on the interval $[t_0, t_0 + h]$. Let us denote

$$m_j(t) = |x_{j+1}(t) - x_j(t)|; \; j = 0, 1, 2, \ldots, \in I^+.$$

Since f satisfies Lipschitz condition and by (5), we have

$$m_j(t) = \left| \int_{t_0}^{t} \left[f(s, x_j(s)) - f(s, x_{j-1}(s)) \right] ds \right|$$
$$\le K \int_{t_0}^{t} \left| x_j(s) - x_{j-1}(s) \right| ds,$$

or, in other words,

$$m_j(t) \le K \int_{t_0}^{t} m_{j-1}(s) ds. \tag{10}$$

By direct computation,

$$m_0(t) = |x_1(t) - x_0(t)| = \left| \int_{t_0}^{t} f(s, x_0(s)) ds \right|$$
$$\le \int_{t_0}^{t} |f(s, x_0(s))| ds$$
$$\le L(t - t_0). \tag{11}$$

We claim that

$$m_j(t) \le LK^j \frac{(t - t_0)^{j+1}}{(j + 1)!}, \tag{12}$$

for j = 0, 1, 2, ..., and $t_0 \le t \le t_0 + h$. The proof of the claim is by induction. For j = 0, (12) is, in fact, (11). Assume that for an integer $1 \le p \le j$ the assertion (12) holds. That is,

$$m_{p+1}(t) \leq K \int_{t_0}^t m_p(s)ds \leq K \int_{t_0}^t LK^p \frac{(s-t_0)^{p+1}}{(p+1)!} ds$$

$$\leq L\, K^{p+1} \frac{(t-t_0)^{p+2}}{(p+2)!}, \quad t_0 \leq t \leq t_0 + h,$$

which shows that (12) holds for j = p + 1 or equivalently, (12) holds for all j≥ o. So,the series $\sum_{j=0}^{\infty} m_j(t)$ is dominated by the series j=o

$$\frac{L}{K} \sum_{j=0}^{\infty} \frac{K^{j+1}h^{j+1}}{(j+1)!},$$

which converges to $\frac{L}{K}\left(e^{Kh}-1\right)$ and hence, the series (9) converges uniformly and absolutely on the $I^+ = [t_0, t_0 + h]$. Let

$$x(t) = x_0(t) + \sum_{n=1}^{\infty} \left[x_n(t) - x_{n-1}(t)\right]; \quad t_0 \leq t \leq t_0 + h. \tag{13}$$

Since the convergence is uniform, the limit function x is continuous on $I^+ = [t_0, t_0 + h]$. Also, the points (t, x(t)) ∈ R for all t ∈ I and thereby completing the proof of (a).

We now show that x satisfies the integral equation

$$x(t) = x_0 + \int_{t_0}^t f(s, x(s))ds, \ t \in I. \tag{14}$$

By the definition of successive approximations

$$x_n(t) = x_0 + \int_{t_0}^t f(s, x_{n-1}(s))ds, \tag{15}$$

from which, we have

$$\left|x(t) - x_0 - \int_{t_0}^t f(s, x(s))ds\right| = \left|x(t) - x_n(t) + \int_{t_0}^t f(s, x_{n-1}(s))ds - \int_{t_0}^t f(s, x(s))ds\right|$$

$$\leq |x(t) - x_n(t)| + \int_{t_0}^t \left|f(s, x_{n-1}(s)) - f(s, x(s))\right|ds. \tag{16}$$

Since $x_n \to x$ uniformly on I, and $|x_n(t) - x_0| \leq b$ for all n and for t ∈ I^+, it follows that $|x(t)| \leq b$ for all t ∈ I^+. Now the Lipschitz condition on f implies

$$|x(t) - x(0) - \int_{t_0}^t f(s, x(s))ds| \leq |x(t) - x_n(t)| + K \int_{t_0}^t |x(s) - x_{n-1}(s)|ds$$

$$\leq |x(t) - x_n(t)| + Kh \max_{t_0 \leq s \leq t_0 + h} |x(s) - x_{n-1}(s)|. \tag{17}$$

The uniform convergence of x_n to x on I^+ now implies that the right hand side of (17) tends to zero as $n \to \infty$. But the left side of (17) is independent of n. Thus, x satisfies the integral equation (6) on I^+ which proves (b).

Uniqueness : Let us now prove that, if $\bar{x}$ and x are any two solutions of the IVP (5), then they co-incide on $[t_0, t_0 + h]$. Let $\bar{x}$ and x satisfy (6) which yields

$$|\bar{x}(t) - x(t)| \leq \int_{t_0}^{t} |f(s, \bar{x}(s)) - f(s, x(s))| ds. \qquad (18)$$

Both $\bar{x}(s)$) and x(s) lie in R for all s in $[t_0, t_0 + h]$ and hence, it follows that

$$|\bar{x}(t) - x(t)| \leq K \int_{t_0}^{t} |\bar{x}(s)) - x(s)| ds.$$

By an application of the Gronwall inequality, we arrive at

$$|\bar{x}(t) - x(t)| \equiv 0 \quad \text{on} \quad [t_0, t_0 + h],$$

which means $\bar{x} \equiv x$. This proves (c), completing the proof of the theorem

Another important feature of Picard's theorem is that a bound for the error (due to truncation of computation at the n-th iteration) can also be obtained. Indeed, we have a result dealing with such a bound on the error.

Corollary.

$$|x(t) - x_n(t)| \leq \frac{L}{K} \frac{(Kh)^{n+1}}{(n+1)!} e^{Kh}; \quad t \in [t_0, t_0 + h]. \qquad (19)$$

Proof. Since

$$x(t) = x_0(t) + \sum_{j=0}^{\infty} \left[x_{j+1}(t) - x_j(t) \right]$$

we have

$$x(t) - x_n(t) = \sum_{j=n}^{\infty} \left[x_{j+1}(t) - x_j(t) \right].$$

Consequently, by (12) we have

$$|x(t) - x_n(t)| \leq \sum_{j=n}^{\infty} |x_{j+1}(t) - x_j(t)| \leq \sum_{j=n}^{\infty} m_j(t) \leq \sum_{j=n}^{\infty} \frac{L}{K} \frac{(Kh)^{j+1}}{(j+1)!}$$

$$= \frac{L}{K} \frac{(Kh)^{n+1}}{(n+1)!} \left[1 + \sum_{j=1}^{\infty} \frac{(Kh)^j}{(n+2)...(n+j+1)} \right]$$

$$\leq \frac{L}{K} \frac{(Kh)^{n+1}}{(n+1)!} e^{Kh}.$$

Continuation and Dependence on Initial Conditions

As usual we assume that the function f in (5) is defined and continuous on an open connected set D and let $(t_0, x_0) \in D$. By Picard's theorem, we have an interval

$$I : t_0 - h \le t \le t_0 + h,$$

where h > 0 such that the closed rectangle $R \subset D$. Since the point $(t_0 + h, x(t_0 + h))$ lies in D there is a rectangle around $(t_0 + h, x(t_0 + h))$ which lies entirely in D. By applying Theorem 1.3.2, we have the existence of a unique solution $\hat{x}$ passing through the point $(t_0 + h, x(t_0 + h))$ and whose graph lies in D (for $t \in \left[t_0 + h, t_0 + h + \hat{h} \right], \hat{h} > 0$). If the solution $\hat{x}$ coincides with x on I, then $\hat{x}$ satisfies the IVP (5) on the interval $[t_0 + h, t_0 + h + \hat{h}] \supset I$. In that case the process may be repeated till the graph of the extended solution reaches the boundary of D. Naturally such a procedure is known as the continuation of solutions of the IVP (5). The continuation method just described can also be extended to the left of t_0.

Now we formalize the above discussion. Let us suppose that a unique solution x of (5) exists, on the interval I* say $h_1 < t < h_2$ with $(t, x(t)) \in D$ for $t \in I^*$ and let

$$|f(t, x)| \le L \text{ on } D, \ (t, x(t)) \in D \text{ and } h_1 < t_0 < h_2.$$

Consider the sequence

$$\left\{ x\left(h_2 - \frac{1}{n}\right) \right\}, n = 1, 2, 3, \ldots.$$

By (6), for sufficiently large n, we have

$$|x(h_2 - \frac{1}{m}) - x(h_2 - \frac{1}{n})| \le \int_{h_2 - (1/n)}^{h_2 - (1/m)} |f(s, x(s))| ds, \quad (m > n)$$

$$\le L |\frac{1}{m} - \frac{1}{n}|.$$

So, the sequence $\left\{ x\left(h_2 - \frac{1}{n}\right) \right\}$ is Cauchy and

$$\lim_{n \to \infty} x\left(h_2 - \frac{1}{n}\right) = \lim_{t \to h_2 - 0} x(t) = x(h_2 - 0),$$

exists. Suppose $\left\{ h_2, x\left(h_2 - 0\right) \right\}$ is in D. Define $\hat{x}$ as follows

$$\hat{x}(t) = x(t), \quad h_1 < t < h_2$$
$$\hat{x}(h_2) = x(h_2 - 0).$$

By noting

$$\hat{x}(t) = x_0 + \int_{t_0}^t f(s, \hat{x}(s))ds, \quad h_1 < t \le h_2,$$

it is easy to show that $\hat{x}$ is a solution of (5) existing on $h_1 < t \le h_2$.

Example : Prove that $\hat{x}$ is a solution of (5) existing on $h_1 < t \le h_2$.

Now consider a rectangle around P : $(h_2, x(h_2 - 0))$ lying inside D. Consider a solutionof (5) through P. As before, by Picard's theorem there exists a solution y through the point P on an interval

$$h_2 - \alpha \le t \le h_2 + \alpha, \; \alpha > 0 \; \text{ and with } \; h_2 - \alpha \ge h_1.$$

Now define z by

$$z(t) = \hat{x}(t), \quad h_1 < t \le h_2$$
$$z(t) = y(t), \quad h_2 \le t \le h_2 + \alpha.$$

Claim: z is a solution of (5) on $h_1 < t \le h_2 + \alpha$. Since y is a unique solution of (5) on $h_2 - \alpha \le t \le h_2 + \alpha$, we have

$$\hat{x}(t) = y(t), \quad h_2 - \alpha \le t \le h_2.$$

We note that z is a solution of (5) on $h_2 \le t \le h_2 + \alpha$ and so it only remains to verify that z' is continuous at the point $t = h_2$. Clearly,

$$z(t) = \hat{x}(h_2) + \int f(s, z(s))ds, \qquad h_2 \le t \le h_2 + \alpha. \tag{20}$$

Further,

$$\hat{x}(h_2) = x_0 + \int_{t_0}^{h_2} f(s, z(s))ds. \tag{21}$$

Thus, the relation (20) and (21) together yield

$$z(t) = x_0 + \int_{t_0}^{h_2} f(s, z(s))ds + \int_{h_2}^t f(s, z(s))ds$$
$$= x_0 + \int_{t_0}^t f(s, z(s))ds, \qquad h_1 \le t \le h_2 + \alpha.$$

Obviously, the derivatives at the end points h_1 and $h_2 + \alpha$ are one-sided. We summarize:

Theorem. Let

(I) $D \subset \mathbb{R}^{n+1}$ be an open connected set and let $f : D \to \mathbb{R}$ be continuous and satisfy the Lipschitz condition in x on D;

(ii) f be bounded on D and

(iii)x be a unique solution of the IVP (5) existing on $h_1 < t < h_2$.

Then,

$$\lim_{t \to h_2 - 0} x(t)$$

exists. If $(h_2, x(h_2 - 0)) \in D$, then x can be continued to the right of h_2.

We now study the continuous dependence of solutions on initial conditions. Consider

$$x' = f(t, x), x(t_0) = x_0. \tag{22}$$

Let $x(t; t_0, x_0)$ be a solution of (22). Then, $x(t; t_0, x_0)$ is a function of time t, the initial time t_0 and the initial state x_0. The dependence on initial conditions means to know about the behavior of $x(t; t_0, x_0)$ as a function of t_0 and x_0. Under certain conditions, indeed x is a continuous function of t_0 and x_0. This amounts to saying that the solution $x(t; t_0, x_0)$ of (22) stays in a neighborhood of solutions $x^*(t; t_0^*, x_0^*)$ of

$$x' = f(t, x), \; x(t_0^*) = x_0^*. \tag{23}$$

provided $\left| t_0 - t_0^* \right|$ and $\left| x_0 - x_0^* \right|$ are sufficiently small. Formally, we have the following theorem:

Theorem. Let $I = [a, b]$ $t_0, t_0^* \in I$ and let $x(t) = x(t; t_0, x_0)$ and $x^*(t) = x(t; t_0^*, x_0^*)$ be solutions of the IVPs (22) and (23) respectively on I. Suppose that (t, x(t)), (t, x*(t)) $\in$ D for t $\in$ I. Further, let f $\in$ Lip(D, K) be bounded by L in D. Then, for any $\epsilon > 0$, there exist a $\delta = \delta(\epsilon) > 0$ such that

$$|x(t) - x^*(t)| < \epsilon, \; t \in I, \tag{24}$$

whenever $\left| t_0 - t_0^* \right| < \delta$ and $\left| x_0 - x_0^* \right| < \delta$.

Proof: It is first of all clear that the solutions x and x* with $x(t_0) = x_0$ and $x^*\left(t_0^*\right) = x_0^*$ exists uniquely. Without loss of generality let $t_0^* \geq t_0$. From Lemma 1.2.1, we have

$$x(t) = x_0 + \int_{t_0}^{t} f(s, x(s))ds, \tag{25}$$

$$x^*(t) = x_0^* + \int_{t_0^*}^{t} f(s, x^*(s))ds. \tag{26}$$

From (25) and (26) we obtain

$$x(t) - x^*(t) = x_0 - x_0^* + \int_{t_0^*}^{t} \left[f(s, x(s)) - f(s, x^*(s)) \right] ds + \int_{t_0}^{t_0^*} f(s, x(s))ds. \tag{27}$$

With absolute values on both sides of (27) and by the hypotheses, we have

$$|x(t) - x^*(t)| \leq |x_0 - x_0^*| + \int_{t_0^*}^t |f(s, x(s)) - f(s, x^*(s))| ds + \int_{t_0}^{t_0^*} |f(s, x(s))| ds$$

$$\leq |x_0 - x_0^*| + \int_{t_0^*}^t K|x(s)) - x^*(s)| ds + L|t_0 - t_0^*|.$$

Now by the Gronwall inequality, it follows that

$$|x(t) - x^*(t)| \leq \left[|x_0 - x_0^*| + L|t_0 - t_0^*|\right] \exp[K(b - a)] \qquad (28)$$

for all t ∈ I. Given any ε > 0, choose

$$\delta(\epsilon) = \frac{\epsilon}{(1 + L) \exp[K(b - a)]}.$$

From (28), we obtain

$$|x(t) - x^*(t)| \leq \delta(1 + L) \exp K(b - a) = \epsilon$$

if $\left|t_0 - t_0^*\right| < \delta(\varepsilon)$ and $\left|x_0 - x_0^*\right| < \delta(\varepsilon)$, which completes the proof.

Remark on Theorems

These theorems clearly exhibit the crucial role played by the Gronwall inequality. Indeed the Gronwall inequality has many more applications in the qualitative theory of differential equations.

Existence of Solutions in the Large

We have seen earlier that the Theorem 1.3.2 is about the existence of solutions in a local sense. In this section, we consider the problem of existence of solutions in the large. Existence of solutions in the large is also known as non-local existence. Before embarking on technical results let us have look at an example.

Example : By Picard's theorem the IVP

$$x' = x^2, \ x(0) = 1, \ -2 \leq t, x \leq 2,$$

has a solution existing on

$$-\frac{1}{2} \leq t \leq \frac{1}{2},$$

where as its solution is

$$x(t) = \frac{1}{1 - t}, \ -\infty < t < 1.$$

Actually, by direct computation, we have an interval of existence larger than the one which we obtain by an application of Picard's theorem. In other words, we need to strengthen the Picard's theorem in order to recover the larger interval of existence.

Now we take up the problem of existence in the large. Under certain restrictions on f, we prove the existence of solutions of IVP

$$x' = f(t, x), \ x(t_0) = x_0, \tag{29}$$

on the whole (of a given finite) interval $|t - t_0| \le T$, and secondly on $-\infty < t < \infty$. We say that x exists "non-locally" on I if x a solution of (29) exists on I. The importance of such problems needs little emphasis due to its necessity in the study of oscillations, stability and boundedness of solutions of IVPs. The non-local existence of solutions of IVP (29) is dealt in the ensuing result.

Theorem. We define a strip S by

$$S = \{(t, x) : |t - t_0| \le T \ and \ |x| < \infty\},$$

where T is some finite positive real number. Let $f : S \to \mathbb{R}$ be a continuous and $f \in \text{Lip}(S, K)$. Then, the successive approximations defined by (7) for the IVP (29) exist on $|t - t_0| \le T$ and converge to a solution x of (29).

Proof: Recall that the definition of successive approximations (7) is

$$\left. \begin{array}{l} x_0(t) \equiv x_0, \\ x_n(t) = x_0 + \displaystyle\int_{t_0}^t f(s, x_{n-1}(s)) ds, \ |t - t_0| \le T. \end{array} \right\} \tag{30}$$

We prove the theorem for the interval $[t_0, t_0 + T]$. The proof for the interval $[t_0 - T, t_0]$ is similar with suitable modifications. First note that (30) defines the successive approximations on $t_0 \le t \le t_0 + T$. Also,

$$|x_1(t) - x_0(t)| = \left| \int_{t_0}^t f(s, x_0(s)) ds \right|. \tag{31}$$

Since f is continuous, $f(t, x_0)$ is continuous on $[t_0, t_0 + T]$ which implies that there exists a real constant L > 0 such that

$$|f(t, x_0)| \le L, \ \text{for all} \ t \in [t_0, t_0 + T].$$

With this bound on $f(t, x_0)$ in (31), we get

$$|x_1(t) - x_0(t)| \le L(t - t_0) \le LT, \ t \in [t_0, t_0 + T]. \tag{32}$$

The estimate (32) implies (by using induction)

$$|x_n(t) - x_{n-1}(t)| \leq \frac{LK^{n-1}T^n}{n!}, \quad t \in [t_0, t_0 + T].$$

(33)

Now (33), as in the proof of Theorem, yields the uniform convergence of the series

$$x_0(t) + \sum_{n=0}^{\infty} [x_{n+1}(t) - x_n(t)],$$

and hence, the uniform convergence of the sequence $\{x_n\}$ on $[t_0, t_0 + T]$ easily follows. Let x denote the limit function, namely,

$$x(t) = x_0(t) + \sum_{n=0}^{\infty} [x_{n+1}(t) - x_n(t)], \quad t \in [t_0, t_0 + T].$$

(34)

In fact, (33) shows that

$$|x_n(t) - x_0(t)| = \left| \sum_{p=1}^{n} [x_p(t) - x_{p-1}(t)] \right|$$

$$\leq \sum_{p=1}^{n} |x_p(t) - x_{p-1}(t)|$$

$$\leq \frac{L}{K} \sum_{p=1}^{n} \frac{K^p T^p}{n!}$$

$$\leq \frac{L}{K} \sum_{p=1}^{\infty} \frac{K^p T^p}{n!} = \frac{L}{K}(e^{KT} - 1).$$

Since x_n converges to x on $t_0 \leq t \leq t_0 + T$, we have

$$|x(t) - x_0| \leq \frac{L}{K}(e^{KT} - 1).$$

Since the function f is continuous on the rectangle

$$R = \{(t, x) : |t - t_0| \leq T, \ |x - x_0| \leq \frac{L}{K}(e^{KT} - 1)\},$$

there exists a real number $L_1 > 0$ such that

$$|f(t, x)| \leq L_1, \ (t, x) \in R.$$

Moreover, the convergence of the sequence $\{x_n\}$ is uniform implies that the limit x is continuous. From the corollary (14), it follows that

$$|x(t) - x_n(t)| \leq \frac{L_1}{K} \frac{(KT)^{n+1}}{(n+1)!} e^{KT}.$$

Finally, we show that x is a solution of the integral equation

$$x(t) = x_0 + \int_{t_0}^{t} f(s, x(s))ds, \quad t_0 \leq t \leq t_0 + T. \tag{35}$$

Also

$$\left|x(t) - x_0 - \int_{t_0}^{t} f(s, x(s))ds\right| = \left|x(t) - x_n(t) + \int_{t_0}^{t} \left[f(s, x_n(s)) - f(s, x(s))\right]ds\right|$$

$$\leq |x(t) - x_n(t)| + \int_{t_0}^{t} \left|f(s, x(t)) - f(s, x_n(s))ds\right| \tag{36}$$

Since $x_n \to x$ uniformly on $[t_0, t_0 + T]$, the right side of (36) tends to zero as $n \to \infty$. By letting $n \to \infty$, from (36) we indeed have

$$\left|x(t) - x_0 - \int_{t_0}^{t} f(s, x(s))ds\right| \leq 0, \quad t \in [t_0, t_0 + T].$$

or else

$$x(t) = x_0 + \int_{t_0}^{t} f(s, x(s))ds, \quad t \in [t_0, t_0 + T].$$

The uniqueness of x follows similarly as shown in the proof of Theorem 1.3.2.

Remark : The example cited at the beginning of this section does not contradict the Theorem 1.5.1 as f (t, x) = x^2 does not satisfy the strip condition f $\in$ Lip(S, K).

A consequence of the Theorem 1.5.1 is:

Theorem. Assume that f (t, x) is a continuous function on $|t| < \infty$, $|x| < \infty$. Further, let f satisfies Lipschitz condition on the the strip S_a for all a > 0, where

$$S_a = \{(t, x) : |t| \leq a, |x| < \infty\}.$$

Then, the initial value problem

$$x' = f(t, x), \quad x(t_0) = x_0, \tag{37}$$

has a unique solution existing for all t.

Proof: The proof is very much based on the fact that for any real number t there exists T such that

$|t - t_0| \le T$. Notice here that all the hypotheses of Theorem are satisfied, for this choice of T, on the strip $|t - t_0| \le T$, $|x| < \infty$. Thus, by Theorem, the successive approximations $\{x_n\}$ converge to a function x which is a unique solution of (37).

Existence and Uniqueness of Solutions of Systems

The methodology developed till now concerns existence and uniqueness of a single equation or usually called a scalar equations which is a natural extension for the study of a system of equations or to higher order equations. In the sequel, we glance at these extensions. Let $I \subseteq \mathbb{R}$ be an interval, $E \subseteq \mathbb{R}^n$. Let $f_1, f_2, ..., f_n : I \times E \to \mathbb{R}$ be given continuous functions.

Consider a system of nonlinear equations

$$
\begin{aligned}
x_1' &= f_1(t, x_1, x_2, ..., x_n), \\
x_2' &= f_2(t, x_1, x_2, ..., x_n), \\
&\cdots\cdots\cdots\cdots\cdots \\
&\cdots\cdots\cdots\cdots\cdots \\
&\cdots\cdots\cdots\cdots\cdots \\
x_n' &= f_n(t, x_1, x_2, ..., x_n),
\end{aligned}
\tag{38}
$$

Denoting (column) vector x with components $x_1, x_2, ..., x_n$ and vector f with components $f_1, f_2, ..., f_n$, the system of equations (38) assumes the form

$$
x' = f(t, x).
\tag{39}
$$

A general n-th order equation is representable in the form (38) which means that the study of n-th order nonlinear equation is naturally embedded in the study of (39). It speaks of the importance of the study of systems of nonlinear equations, leaving apart numerous difficulties that one has to face. Consider an IVP

$$
x' = f(t, x), \quad x(t_0) = x_0.
\tag{40}
$$

The proofs of local and non-local existence theorems for systems of equations stated below have a remarkable resemblance to those of scalar equations. The detailed proofs are to be supplied by readers with suitable modifications to handle the presence of vectors and their norms. Below the symbol |.| is used to denote both the norms of a vector and the absolute value. There is no possibility of confusion since the context clarifies the situation.

In all of what follows we are concerned with the region D, a rectangle in $\mathbb{R}^{n+1}$ space, defined by

$$
D = \{(t, x) : |t - t_0| \le a, |x - x_0| \le b\},
$$

where x, $x_0 \in \mathbb{R}^n$ and t, $t_0 \in \mathbb{R}$.

Definition. A function $f : D \to \mathbb{R}^n$ is said to satisfy the Lipschitz condition in the variable x, with Lipschitz constant K on D if

$$|f(t, x_1) - f(t, x_2)| \leq K|x_1 - x_2|$$

(41)

uniformly in t for all (t, x_1), (t, x_2) in D.

The continuity of f in x for each fixed t is a consequence, when f is Lipschitzian in x. If f is Lipschitzian on D then, there exists a non-negative, real-valued function L(t) such that

$$|f(t, x)| \leq L(t), \text{ for all } (t, x) \in D.$$

In addition, there exists a constant L > 0 such that L(t) ≤ L, when L is continuous on $|t - t_0| \leq a$. We note that L depends on f and many write L_f instead of L to denotes its dependence on f.

Lemma. Let $f : D \to \mathbb{R}^n$ be a continuous function. $x(t; t_0, x_0)$ (denoted by x) is a solution of (40) on some interval I contained in $|t - t_0| \leq a(t_0 \in I)$ if and only if x is a solution of the integral equation

$$x(t) = x_0 + \int_{t_0}^{t} f(s, x(s))ds, \ t \in I.$$

(42)

Proof: First of all, we prove that the components x_i of x satisfy

$$x_i(t) = x_{0i} + \int_{t_0}^{t} f_i(s, x(s))ds, \quad t \in I, \ i = 1, 2, \ldots, n,$$

if and only if

$$x_i'(t) = f_i(t, x(t)), \ x_{0i} = x_i(t_0), \quad i = 1, 2, \ldots, n,$$

holds. The proof is exactly the same as that of Lemma 1.2.1 and hence omitted.

As expected, the integral equation (42) is now exploited to define (inductively) the successive approximations by

$$\begin{cases} x_0(t) = x_0 \\ x_n(t) = x_0 + \int_{t_0}^{t} f(s, x_{n-1}(s))ds, \quad t \in I \end{cases}$$

(43)

for n = 1, 2, . . . , . The ensuing lemma establishes that, under the stated conditions, the successive approximations are indeed well defined.

Lemma. Let $f : D \to \mathbb{R}^n$ be a continuous function and be bounded by L > 0 on D.

Define $h = \min\left(a, \dfrac{b}{L}\right)$. Then, the successive approximations are well defined by (43) on the interval $I = |t - t_0| \leq h$. Further,

$$|x_j(t) - x_0| \le L\,|t - t_0| < b, \quad j = 1, 2, \ldots$$

Theorem. (Picard's theorem for system of equations). Let all the conditions of Lemma hold and let f satisfy the Lipschitz condition with Lipschitz constant K on D. Then, the successive approximations defined by (43) converge uniformly on I = |t−t$_o$| ≤ h to a unique solution of the IVP (40).

Corollary. A bound error left due to the truncation at the n-th approximation for x is

$$|x(t) - x_n(t)| \le \frac{L}{K}\frac{(Kh)^{n+1}}{(n+1)!}e^{Kh}, \quad t \in [t_0, t_0 + h].$$

$$(44)$$

Corollary. Let $M_n(\mathbb{R})$ denote the set of all n × n real matrices. Let $I \subset \mathbb{R}$ be an interval. Let $A : I \to \mathbb{R}$ be continuous on I. Then, the IVP

$$x' = A(t)x,$$
$$x(a) = x_0, \ a \in I,$$

has a unique solution x existing on I. As a consequence the set of all solutions of

$$x' = Ax,$$

is a linear vector space of dimension n.

As noted earlier the Lipschitz property of f in Theorem 1.6.4 cannot be altogether dropped as shown by the following example.

Example. The nonlinear IVP

$$x_1' = 2x_2^{1/3}, \quad x_1(0) = 0,$$
$$x_2' = 3x_1, \quad x_2(0) = 0,$$

in the vector form is

$$x' = f(t, x), \quad x(0) = \mathbf{0},$$

where x = (x$_1$, x$_2$), $f(t, x) = \left(2x_2^{1/3}, 3x_1\right)$ and 0 is the zero vector. Obviously, x(t) ≡ 0 is a solution. It is easy to verify that x(t) = (t², t³) is yet another solution of the IVP which violates the uniqueness of the solutions of IVP .

References

- Liao, S.J. (2003), Beyond Perturbation: Introduction to the Homotopy Analysis Method, Boca Raton: Chapman & Hall/ CRC Press, ISBN 1-58488-407-X

- Lewy, Hans (1957), "An example of a smooth linear partial differential equation without solution", Annals of

Mathematics. Second Series, 66 (1): 155–158, doi:10.2307/1970121

- Donchev, Tzanko; Farkhi, Elza (1998). "Stability and Euler Approximation of One-sided Lipschitz Differential Inclusions". SIAM Journal on Control and Optimization. 36 (2): 780–796. doi:10.1137/S0363012995293694

- Roubíček, T. (2013), Nonlinear Partial Differential Equations with Applications (2nd ed.), Basel, Boston, Berlin: Birkhäuser, ISBN 978-3-0348-0512-4, MR MR3014456

- Bellman, Richard (1943), "The stability of solutions of linear differential equations", Duke Math. J., 10 (4): 643–647, MR 0009408, Zbl 0061.18502, doi:10.1215/s0012-7094-43-01059-2

- Krasil'shchik, I.S. & Vinogradov, A.M., Eds. (1999), Symmetries and Conservation Laws for Differential Equations of Mathematical Physics, American Mathematical Society, Providence, Rhode Island,USA, ISBN 0-8218-0958-X

2

Partial Differential Equations of First-Order

First order partial differential equations help in understanding several problems concerned with science and technology. These equations are equations that are concerned with first derivatives. They are used in calculus of variations, in the construction of characteristic surfaces for hyperbolic partial differential equations, etc. This chapter will provide an integrated understanding of first-order partial differential equations.

First-order Partial Differential Equation

In mathematics, a first-order partial differential equation is a partial differential equation that involves only first derivatives of the unknown function of n variables. The equation takes the form

$$F(x_1,\ldots,x_n,u,u_{x_1},\ldots u_{x_n}) = 0.$$

Such equations arise in the construction of characteristic surfaces for hyperbolic partial differential equations, in the calculus of variations, in some geometrical problems, and in simple models for gas dynamics whose solution involves the method of characteristics. If a family of solutions of a single first-order partial differential equation can be found, then additional solutions may be obtained by forming envelopes of solutions in that family. In a related procedure, general solutions may be obtained by integrating families of ordinary differential equations.

Characteristic Surfaces for the Wave Equation

Characteristic surfaces for the wave equation are level surfaces for solutions of the equation

$$u_t^2 = c^2 \left(u_x^2 + u_y^2 + u_z^2 \right).$$

There is little loss of generality if we set $u_t = 1$: in that case u satisfies

$$u_x^2 + u_y^2 + u_z^2 = \frac{1}{c^2}.$$

In vector notation, let

$$\vec{x} = (x,y,z) \quad \text{and} \quad \vec{p} = (u_x,u_y,u_z).$$

A family of solutions with planes as level surfaces is given by

$$u(\vec{x}) = \vec{p} \cdot (\vec{x} - \vec{x_0}),$$

where

$$|\vec{p}| = \frac{1}{c}, \quad \text{and} \quad \vec{x_0} \quad \text{is arbitrary.}$$

If x and x_0 are held fixed, the envelope of these solutions is obtained by finding a point on the sphere of radius $1/c$ where the value of u is stationary. This is true if $\vec{p}$ is parallel to $\vec{x} - \vec{x_0}$. Hence the envelope has equation

$$u(\vec{x}) = \pm \frac{1}{c} | \vec{x} - \vec{x_0} |.$$

These solutions correspond to spheres whose radius grows or shrinks with velocity c. These are light cones in space-time.

The initial value problem for this equation consists in specifying a level surface S where $u=0$ for $t=0$. The solution is obtained by taking the envelope of all the spheres with centers on S, whose radii grow with velocity c. This envelope is obtained by requiring that

$$\frac{1}{c} | \vec{x} - \vec{x_0} | \quad \text{is stationary for} \quad \vec{x_0} \in S.$$

This condition will be satisfied if $| \vec{x} - \vec{x_0} |$ is normal to S. Thus the envelope corresponds to motion with velocity c along each normal to S. This is the **Huygens' construction of wave fronts**: each point on S emits a spherical wave at time $t=0$, and the wave front at a later time t is the envelope of these spherical waves. The normals to S are the light rays.

Two-dimensional Theory

The notation is relatively simple in two space dimensions, but the main ideas generalize to higher dimensions. A general first-order partial differential equation has the form

$$F(x, y, u, p, q) = 0,$$

where

$$p = u_x, \quad q = u_y.$$

A complete integral of this equation is a solution $\phi(x,y,u)$ that depends upon two parameters a and b. (There are n parameters required in the n-dimensional case.) An envelope of such solutions is obtained by choosing an arbitrary function w, setting $b=w(a)$, and determining $A(x,y,u)$ by requiring that the total derivative

$$\frac{d\varphi}{da} = \varphi_a(x, y, u, A, w(A)) + w'(A)\varphi_b(x, y, u, A, w(A)) = 0.$$

In that case, a solution u_w is also given by

$$u_w = \phi(x, y, u, A, w(A))$$

Each choice of the function w leads to a solution of the PDE. A similar process led to the construction of the light cone as a characteristic surface for the wave equation.

If a complete integral is not available, solutions may still be obtained by solving a system of ordinary equations. To obtain this system, first note that the PDE determines a cone (analogous to the light cone) at each point: if the PDE is linear in the derivatives of u (it is quasi-linear), then the cone degenerates into a line. In the general case, the pairs (p,q) that satisfy the equation determine a family of planes at a given point:

$$u - u_0 = p(x - x_0) + q(y - y_0),$$

where

$$F(x_0, y_0, u_0, p, q) = 0.$$

The envelope of these planes is a cone, or a line if the PDE is quasi-linear. The condition for an envelope is

$$F_p \, dp + F_q \, dq = 0,$$

where F is evaluated at (x_0, y_0, u_0, p, q), and dp and dq are increments of p and q that satisfy $F=0$. Hence the generator of the cone is a line with direction

$$dx : dy : du = F_p : F_q : (pF_p + qF_q).$$

This direction corresponds to the light rays for the wave equation. To integrate differential equations along these directions, we require increments for p and q along the ray. This can be obtained by differentiating the PDE:

$$F_x + F_u p + F_p p_x + F_q p_y = 0,$$

$$F_y + F_u q + F_p q_x + F_q q_y = 0,$$

Therefore the ray direction in (x, y, u, p, q) space is

$$x : dy : du : dp : dq = F_p : F_q : (pF_p + qF_q) : (-F_x - F_u p) : (-F_y - F_u q).$$

The integration of these equations leads to a ray conoid at each point (x_0, y_0, u_0). General solutions of the PDE can then be obtained from envelopes of such conoids.

A first order PDE in two independent variables x, y and the dependent variable z can be written in the form

$$f\left(x, y, z, \frac{\partial z}{\partial x}, \frac{\partial z}{\partial y}\right) = 0 \qquad\qquad (1)$$

For convenience, we set

$$p = \frac{\partial z}{\partial x}, \quad q = \frac{\partial z}{\partial y}.$$

Equation (1) then takes the form

$$f(x, y, z, p, q) = 0 \qquad\qquad (2)$$

The equations of the type (2) arise in many applications in geometry and physics. For instance, consider the following geometrical problem.

EXAMPLE 1. Find all functions $z(x, y)$ such that the tangent plane to the graph $z = z(x, y)$ at any arbitrary point $(x_0, y_0, z(x_0, y_0))$ passes through the origin characterized by the PDE $xz_x + yz_y - z = 0$.

The equation of the tangent plane to the graph at $(x_0, y_0, z(x_0, y_0))$ is

$$z_x(x_0, y_0)(x - x_0) + z_y(x_0, y_0)(y - y_0) - (z - z(x_0, y_0)) = 0$$

This plane passes through the origin $(0,0,0)$ and hence, we must have

$$-z_x(x_0, y_0)x_0 - z_y(x_0, y_0)y_0 + z(x_0, y_0) = 0 \qquad\qquad (3)$$

For the equation (3) to hold for all (x_0, y_0) in the domain of z, z must satisfy

$$xz_x + yz_y - z = 0,$$

which is a first-order PDE.

EXAMPLE 2. The set of all spheres with centers on the z-axis is characterized by the first-order PDE $yp - xq = 0$.

The equation

$$x^2 + y^2 + (z - c)^2 = r^2, \qquad\qquad (4)$$

where r and c are arbitrary constants, represents the set of all spheres whose centers lie on the z-axis. Differentiating (4) with respect to x, we obtain

$$2\left(x + (z - c)\frac{\partial z}{\partial x}\right) = 2(x + (z - c)p) = 0 \qquad\qquad (5)$$

Differentiate (4) with respect to y to have

$$y + (z - c)q = 0 \qquad\qquad (6)$$

Eliminating the arbitrary constant c from (5) and (6), we obtain the first-order PDE

$$yp - xq = 0 \qquad (7)$$

Equation (4) in some sense characterized the first-order PDE (7).

EXAMPLE 3. Consider all surfaces described by an equation of the form

$$z = f\left(x^2 + y^2\right), \qquad (8)$$

where f is an arbitrary function, described by the first-order PDE.

Writing $u = x^2 + y^2$ and differentiating (8) with respect to x and y, it follows that

$$p = 2xf'(u); \quad q = 2yf'(u),$$

where $f'(u) = \dfrac{df}{du}$. Eliminating f '(u) from the above two equations, we obtain the same first-order PDE as in (7).

REMARK 4. The function z described by each of the equations (4) and (8), in some sense, a solution to the PDE (7). Observe that, in Example 2, PDE (7) is formulated by eliminating arbitrary constants from (4) whereas in Example 3, PDE (7) is formed by eliminating an arbitrary function.

Formation of First-order PDEs

The applications of conservation principles often yield a first-order PDEs. We have seen in the previous two examples that a first-order PDE can be formed either by eliminating arbitrary constants or an arbitrary function involved. Below, we now generalize the arguments of Example 2 and Example 3 to show that how a first-order PDE can be formed.

Method I (Eliminating arbitrary constants): Consider two parameters family of surfaces described by the equation

$$f(x, y, z, a, b) = 0 \qquad (9)$$

where a and b are arbitrary constants. Equation (9) may be thought of as a generalization of the relation (4).

Differentiating (9) with respect to x and y, we obtain

$$\frac{\partial F}{\partial x} + p\frac{\partial F}{\partial z} = 0 \qquad (10)$$

$$\frac{\partial F}{\partial y} + q\frac{\partial F}{\partial z} = 0 \qquad (11)$$

Eliminate the constants a, b from equations (9), (10) and (11) to obtain a first-order PDE of the for

$$f(x, y, z, p, q) = 0 \qquad (12)$$

This shows that a family of surfaces described by the relation (9) gives rise to a first-order PDE (12).

Method II (Eliminating arbitrary function): Now consider the generalization of Example 3. Let

$u(x, y, z) = c_1$ and $v(x, y, z) = c_2$ be two known functions of x, y and z satisfying a relation of the form

$$F(u,v) = 0 \qquad (13)$$

where F is an arbitrary function of u and v. Differentiating (13) with respect to x and y lead to the equations

$$F_u(u_x + u_z p) + F_v(v_x + v_z p) = 0$$
$$F_u(u_y + u_z q) + F_v(v_y + v_z q) = 0$$

Eliminating F_u and F_v from the above two equations, we obtain

$$p\frac{\partial(u,v)}{\partial(y,z)} + q\frac{\partial(u,v)}{\partial(z,x)} = \frac{\partial(u,v)}{\partial(x,y)} \qquad (14)$$

which is a first-order PDE of the form $f(x,y,z,p,q) = 0$. Here, $\dfrac{\partial(u,v)}{\partial(x,y)} = u_x v_y - u_y v_x$.

Classification of First-order PDEs

We classify the equation (1) depending on the special forms of the function f. If (1) is of the form

$$a(x,y)\frac{\partial z}{\partial x} + b(x,y)\frac{\partial z}{\partial y} + c(x,y)z = d(x,y)$$

then it is called linear first-order PDE. Note that the function f is linear in $\dfrac{\partial z}{\partial x}, \dfrac{\partial z}{\partial y}$ and z with all coefficients depending on the independent variables x and y only.

If (1) has the form

$$a(x,y)\frac{\partial z}{\partial x} + b(x,y)\frac{\partial z}{\partial y} = c(x,y,z)$$

then it is called semilinear because it is linear in the leading (highest-order) terms $\dfrac{\partial z}{\partial x}$ and $\dfrac{\partial z}{\partial y}$. However, it need not be linear in z. Note that the coefficients of $\dfrac{\partial z}{\partial x}$ are $\dfrac{\partial z}{\partial y}$ are functions of the independent variables only.

If (1) has the form

$$a(x,y,z)\frac{\partial z}{\partial x} + b(x,y,z)\frac{\partial z}{\partial y} = c(x,y,z)$$

then it is called quasi-linear PDE. Here the function f is linear in the derivatives $\dfrac{\partial z}{\partial x}$ and $\dfrac{\partial z}{\partial y}$ with the coefficients a, b and c depending on the independent variables x and y as well as on the unknown z. Note that linear and semilinear equations are special cases of quasi-linear equations.

Any equation that does not fit into one of these forms is called nonlinear.

EXAMPLE

$$1.\ xz_x + yz_y = z\ (linear)$$

$$2.\ xz_x + yz_y = z^2\ (semilinear)$$

$$3.\ z_x + (x+y)z_y = xy\ (linear)$$

$$4.\ zz_x + z_y = 0\ (quasilinear)$$

$$5.\ xz_x^2 + yz_y^2 = 2(nonlinear)$$

Cauchy's Problem or IVP for First-order PDEs

Recall the initial value problem for a first-order ODE which ask for a solution of the equation that takes a given value at a given point of $\mathbb{R}$. The IVP for first-order PDE ask for a solution of (2) which has given values on a curve in $\mathbb{R}^2$. The conditions to be satisfied in the case of IVP for first-order PDE are formulated in the classic problem of Cauchy which may be stated as follows:

Let C be a given curve in $\mathbb{R}$ described parametrically by the equations

$$x = x_0(s),\ y = y_0(s)\ ;\ s \in I, \tag{15}$$

where $x_0(s)$, $y_0(s)$ are in $C^1(I)$. Let $z_0(s)$ be a given function in $C^1(I)$. The IVP or Cauchy's problem for first-order PDE

$$f(x,y,z,p,q) = 0 \tag{16}$$

is to find a function $u = u(x,\ y)$ with the following properties:

- $u(x,\ y)$ and its partial derivatives with respect to x and y are continuous in a region Ω of $\mathbb{R}^2$ containing the curve C.

- $u = u(x,\ y)$ is a solution of (16) in Ω, i.e.,

$$f(x,y,u(x,y),u_x(x,y),u_y(x,y)) = 0\ in\ \Omega$$

- On the curve C

$$u(x_0(s), y_0(s)) = z_0(s),\ s \in I \tag{17}$$

The curve C is called the initial curve of the problem and the function $z_0(s)$ is called the initial data. Equation (17) is called the initial condition of the problem.

NOTE: Geometrically, Cauchy's problem may be interpreted as follows: To find a solution surface $u = u(x,\ y)$ of (16) which passes through the curve C whose parametric equations are

$$x = x_0(s),\ y = y_0(s)\ z = z_0(s) \tag{18}$$

Further, at every point of which the direction $(p,q,-1)$ of the normal is such that

$$f(x,y,z,p,q) = 0$$

The proof of existence of a solution of (16) passing through a curve with equations (18) requires some more assumptions on the function f and the nature of the curve C. We now state the classic theorem due to Kowalewski in the following theorem.

THEOREM. (Kowalewski) If $g(y)$ and all its derivatives are continuous for $|y - y_0| < \delta$, if x_0 is a given number and $z_0 = g(y_0), q_0 = g'(y_0)$, and if $f(x, y, z, q)$ and all its partial derivatives are continuous in a region S defined by

$$|x - x_0| < \delta, |y - y_0| < \delta, |q - q_0| < \delta$$

then there exists a unique function $\phi(x, y)$ such that:

(a) $\phi(x, y)$ and all its partial derivatives are continuous in a region

$$\Omega : |x - x_0| < \delta_1, |y - y_0| < \delta_2;$$

(b) For all (x, y) in Ω, $z = \phi(x, y)$ is a solution of the equation

$$\frac{\partial z}{\partial x} = f(x, y, z, \frac{\partial z}{\partial y})$$

(c) For all values of y in the interval $|y - y_0| < \delta_1, \phi(x_0, y) = g(y)$.

DEFINITION. (A complete solution or a complete integral) Any relation of the form

$$F(x, y, z, a, b) = 0 \qquad\qquad (19)$$

which contains two arbitrary constants a and b and is a solution of a first-order PDE is called a complete solution or a complete integral of that first-order PDE.

DEFINITION. (A general solution or a general integral) Any relation of the form

$$F(u, v) = 0$$

involving an arbitrary function F connecting two known functions $u(x, y, z)$ and $v(x, y, z)$ and providing a solution of a first-order PDE is called a general solution or a general integral of that first-order PDE.

It is possible to derive a general integral of the PDE once a complete integral is known.

With $b = \phi(a)$, if we take any one-parameter subsystem

$$f(x, y, z, a, \phi(a)) = 0$$

of the system (19) and form its envelope, we obtain a solution of equation (16). When $\phi(a)$ is arbitrary, the solution obtained is called the general integral of (16) corresponding to the complete integral (19).

When a definite $\phi(a)$ is used, we obtain a particular solution.

DEFINITION. (A singular integral) The envelope of the two-parameter system (19) is also a solution of the equation (16). It is called the singular integral or singular solution of the equation.

Linear First-Order Partial Differential Equations

The most general first-order linear PDE has the form

$$a(x,y)z_x + b(x,y)z_y + c(x,y)z = d(x,y), \tag{1}$$

where a, b, c, and d are given functions of x and y. These functions are assumed to be continuously differentiable. Rewriting (1) as

$$a(x,y)z_x + b(x,y)z_y = -c(x,y)z + d(x,y) \tag{2}$$

we observe that the left hand side of (2), i.e.,

$$a(x,y)z_x + b(x,y)z_y = \nabla z \cdot (a,b)$$

is (essentially) a directional derivative of $z(x,y)$ in the direction of the vector (a,b), where (a,b) is defined and nonzero. When a and b are constants, the vector (a,b) had a fixed direction and magnitude, but now the vector can change as its base point (x,y) varies. Thus, (a,b) is a vector field on the plane.

The equations

$$\frac{dx}{dt} = a(x,y), \frac{dy}{dt} = b(x,y), \tag{3}$$

determine a family of curves $x = x(t)$, $y = y(t)$ whose tangent vector $(\frac{dx}{dt}, \frac{dy}{dt})$ coincides with the direction of the vector (a,b). Therefore, the derivative of $z(x,y)$ along these curves becomes

$$\begin{aligned}
\frac{dz}{dt} = \frac{d}{dt}z\{(x(t),y(t))\} &= \frac{\partial z}{\partial x}\frac{dx}{dt} + \frac{\partial z}{\partial y}\frac{dy}{dt} \\
&= z_x(x(t),y(t))a(x(t),y(t)) + z_y(x(t),y(t))b(x(t),y(t)) \\
&= -c(x(t),y(t))z(x(t),y(t)) + d(x(t),y(t)) \\
&= -c(t)z(t) + d(t),
\end{aligned}$$

where we have used the chain rule and (1). Thus, along these curves, $z(t) = z(x(t),y(t))$ satisfies the ODE

$$z'(t) + c(t)z(t) = d(t). \tag{4}$$

Let $\mu(t) = \exp\left[\int_0^t c(\tau)d\tau\right]$ be an integrating factor for (4). Then, the solution is given by

$$z(t) = \frac{1}{\mu(t)}\left[\int_0^t \mu(\tau)d(\tau)d\tau + z(0)\right]. \tag{5}$$

The approach described above to solve (1) by using the solutions of (3)-(4) is called the method of characteristics. It is based on the geometric interpretation of the partial differential equation (1).

NOTE: (i) The ODEs (3) is known as the characteristics equation for the PDE (1). The solution curves of the characteristic equation are the characteristics curves for (1).

(ii) Observe that $\mu(t)$ and $d(t)$ depend only on the values of $c(x, y)$ and $d(x, y)$ along the characteristics curve $x = x(t)$, $y = y(t)$. Thus, equation (5) shows that the values $z(t)$ of the solution z along the entire characteristics curve are completely determined, once the value $z(0) = z(x(0), y(0))$ is prescribed.

(iii) Assuming certain smoothness conditions on the functions a, b, c, and d, the existence and uniqueness theory for ODEs guarantees a unique solution curve $(x(t), y(t), z(t))$ of (3)-(4) (i.e., a characteristic curve) passes through a given point (x_0, y_0, z_0) in (x, y, z)- space.

The Method of Characteristics for Solving Linear First-order IVP

In practice we are not interested in determining a general solution of the partial differential equation (1) but rather a specific solution $z = z(x, y)$ that passes through or contains a given curve C. This problem is known as the initial value problem for (1). The method of characteristics for solving the initial value problem for (1) proceeds as follows.

Let the initial curve C be given parametrically as:

$$x = x(s), \quad y = y(s), \quad z = z(s). \qquad (6)$$

for a given range of values of the parameter s. The curve may be of finite or infinite extent and is required to have a continuous tangent vector at each point.

Every value of s fixes a point on C through which a unique characteristic curve passes. The family of characteristic curves determined by the points of C may be parameterized as

$$x = x(s,t), \quad y = y(s,t), \quad z = z(s,t)$$

with $t = 0$ corresponding to the initial curve C. That is, we have

$$x(s,0) = x(s), \quad y(s,0) = y(s), \quad z(s,0) = z(s).$$

In other words, we have the following:

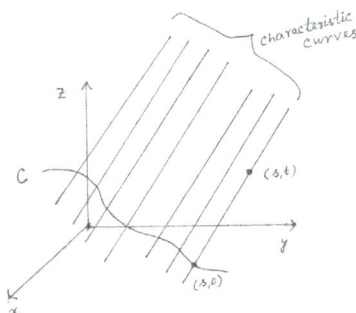

Characteristic curves and construction of the integral surface

The functions x(s, t) and y(s, t) are the solutions of the characteristics system (for each fixed s)

$$\frac{d}{dt}x(s,t) = a(x(s,t), y(s,t)), \quad \frac{d}{dt}y(s,t) = b(x(s,t), y(s,t))$$

(7)

with given initial values x(s, 0) and y(s, 0).

Suppose that

$$z(x(s,0), y(s,0)) = g(s),$$

(8)

where $g(s)$ is a given function. We obtain $z(x(s,t), y(s,t))$ as follows: Let

$$z(s,t) = z(x(s,t), y(s,t)), c(s,t) = c(x(s,t), y(s,t)), d(s,t) = d(x(s,t), y(s,t))$$

(9)

and

$$\mu(s,t) = \exp\left[\int_0^t c(s,t)dt\right]$$

(10)

Analogous to formula (5), for each fixed s, we obtain

$$z(s,t) = \frac{1}{(s,t)}\left[\int (s,t)d(s,t)dt + g(s)\right]$$

(11)

$z(s,t)$ is the value of z at the point $(x(s,t), y(s,t))$. Thus, as s and t vary, the point (x, y, z), in xyz-space, given by

$$x = x(s,t), \ y = y(s,t), \ z = z(s,t)$$

(12)

traces out the surface of the graph of the solution z of the PDE (1) which meets the initial curve (8). The equations (12) constitute the parametric form of the solution of (1) satisfying the initial condition (8) [i.e., a surface in (x, y, z)-space that contains the initial curve].

NOTE: If the Jacobian $J(s,t) = x_s y_t - x_t y_s \neq 0$, then the equations $x = x(s,t)$ and $y = y(s,t)$ can be inverted to give s and t as (smooth) functions of x and y i.e., $s = s(x, y)$ and $t = t(x, y)$. The resulting function $z = z(x, y) = z(s(x,y), t(x, y))$ satisfies the PDE(1) in a neighborhood of the curve C (in view of (4) and the initial condition (6)) and is the unique solution of the IVP.

EXAMPLE. Determine the solution the following IVP:

$$\frac{\partial z}{\partial y} + c\frac{\partial z}{\partial x} = 0, \ z(x,0) = f(x)$$

where $f(x)$ is a given function and c is a constant.

Solution: A step by step procedure for the finding solution is given below.

Step 1:(Finding characteristic curves)

To apply the method of characteristics, parameterize the initial curve C as follows: as follows:

$$x = s, \, y = 0, \, z = f(s) \tag{13}$$

The family of characteristics curves $x\big((s,t), y(s,t)\big)$ are determined by solving the ODEs

$$\frac{d}{dt} x(s,t) = c, \, \frac{d}{dt} y(s,t) = 1$$

The solution of the system is

$$x(s,t) = ct + c_1(s) \; \text{ and } \; y(s,t) = t + c_2(s)$$

Step 2: (Applying IC) Using the initial conditions

$$x(s,0) = s \, , \, y(s,0) = 0$$

we find that

$$c_1(s) = s, \, c_2(s) = 0$$

and hence

$$x(s,t) = ct + s \; \text{ and } \; y(s,t) = t$$

Step 3: (Writing the parametric form of the solution)

Comparing with (1), we have $c(x,y) = 0$ and $d(x,y) = 0$. Therefore, using (10) and (11), we find that

$$d(s,t) = 0, \, \mu(s,t) = 1$$

Since $z\big(x(s,0), y(s,0)\big) = z(s,0) = g(s) = f(s)$, we obtain $z(s,t) = f(s)$. Thus, the parametric form of the solution of the problem is given by

$$x(s,t) = ct + s, \; y(s,t) = t, \, z(s,t) = f(s)_.$$

Step 4: (Expressing $z(s,t)$ in terms of $z(x,y)$) Expressing s and t as $s = s(x,y)$ and $t = t(x,y)$, we have

$$s = x - cy, \; t = y.$$

We now write the solution in the explicit form as

$$z(x,y) = z(s(x,y), y(x,y)) = f(x - cy).$$

Clearly, if $f(x)$ is differentiable, the solution $z(x,y) = f(x - cy)$ satisfies given PDE as well as the initial condition.

NOTE: Example 1 characterizes unidirectional wave motion with velocity c. If we consider the initial function $z(x,0) = f(x)$ to represent a waveform, the solution $z(x,y) = f(x-cy)$ shows that a point x for which $x - cy = $ constant , will always occupy the same position on the wave form. If $c > 0$, the entire initial wave form $f(x)$ moves to the right without changing its shape with speed c (if $c < 0$, the direction of motion is reversed).

EXAMPLE. Find the parametric form of the solution of the problem

$$-yz_x + xz_y = 0$$

with the condition given by

$$z(s, s^2) = s^3, (s < 0).$$

Solution: To find the solution, let's proceed as follows.

Step 1: (Finding characteristic curves)

The family of characteristics curves $(x(s,t), y(s,t))$ are determined by solving

$$\frac{d}{dt}x(s,t) = -y(s,t), \frac{d}{dt}y(s,t) = x(s,t)$$

with initial conditions

$$x(s,0) = s, \ y(s,0) = s^2$$

The general solution of the system is

$$x(s,t) = c_1(s)\cos(t) + c_2(s)\sin(t) \quad and \ y(s,t) = c_1(s)\sin(t) - c_2(s)\cos(t)$$

Step 2: (Applying IC)

Using ICs, we find that

$$c_1(s) = s \ , c_2(s) = -s^2$$

and hence

$$x(s,t) = s\cos(t) - s^2 \sin(t) \quad and \ y(s,t) = s\sin(t) + s^2 \cos(t)$$

Step 3: (Writing the parametric form of the solution)

Comparing with (1), we note that $c(x,y) = 0$ and $d(x,y) = 0$. Therefore, using (10) and (11), it follows that

$$d(s,t) = 0, \quad \mu(s,t) = 1.$$

In view of the given condition curve and $z = z(s,t)$, we obtain

$$z(x(s,0), y(s,0)) = z(s,s^2) = g(s) = s^3, z(s,t) = s^3.$$

Thus, the parametric form of the solution of the problem is given by

$$x(s,t) = s\cos(t) - s^2\sin(t), \ y(s,t) = s\sin(t) + s^2\cos(t), \ z(s,t) = s^3.$$

Step 4: (Expressing $z(s,t)$ in terms of $z(x,y)$)

Writing s and t as a function of x and y, it is an easy exercise to show that

$$z(x,y) = \frac{1}{\sqrt{8}}\left[-1 + \sqrt{1 + 4(x^2 + y^2)}\right]^{3/2}.$$

Method of Characteristics

In mathematics, the method of characteristics is a technique for solving partial differential equations. Typically, it applies to first-order equations, although more generally the method of characteristics is valid for any hyperbolic partial differential equation. The method is to reduce a partial differential equation to a family of ordinary differential equations along which the solution can be integrated from some initial data given on a suitable hypersurface.

Characteristics of First-order Partial Differential Equation

For a first-order PDE (partial differential equation), the method of characteristics discovers curves (called characteristic curves or just characteristics) along which the PDE becomes an ordinary differential equation (ODE). Once the ODE is found, it can be solved along the characteristic curves and transformed into a solution for the original PDE.

For the sake of motivation, we confine our attention to the case of a function of two independent variables x and y for the moment. Consider a quasilinear PDE of the form

$$a(x,y,z)\frac{\partial z}{\partial x} + b(x,y,z)\frac{\partial z}{\partial y} = c(x,y,z). \qquad (1)$$

Suppose that a solution z is known, and consider the surface graph $z = z(x,y)$ in R³. A normal vector to this surface is given by

$$\left(\frac{\partial z}{\partial x}(x,y), \frac{\partial z}{\partial y}(x,y), -1\right).$$

As a result, equation (1) is equivalent to the geometrical statement that the vector field

$$(a(x,y,z), b(x,y,z), c(x,y,z))$$

is tangent to the surface $z = z(x,y)$ at every point, for the dot product of this vector field with the above normal vector is zero. In other words, the graph of the solution must be a union of integral curves of this vector field. These integral curves are called the characteristic curves of the original partial differential equation.

The equations of the characteristic curve may be expressed invariantly by the *Lagrange-Charpit equations*

$$\frac{dx}{a(x,y,z)} = \frac{dy}{b(x,y,z)} = \frac{dz}{c(x,y,z)},$$

or, if a particular parametrization t of the curves is fixed, then these equations may be written as a system of ordinary differential equations for $x(t)$, $y(t)$, $z(t)$:

$$\frac{dx}{dt} \quad a(x,y,z)$$

$$\frac{dy}{dt} \quad b(x,y,z)$$

$$\frac{dz}{dt} \quad c(x,y,z).$$

These are the characteristic equations for the original system.

Linear and Quasilinear Cases

Consider now a PDE of the form

$$\sum_{i=1}^{n} a_i(x_1,\ldots,x_n,u)\frac{\partial u}{\partial x_i} = c(x_1,\ldots,x_n,u).$$

For this PDE to be linear, the coefficients a_i may be functions of the spatial variables only, and independent of u. For it to be quasilinear, a_i may also depend on the value of the function, but not on any derivatives. The distinction between these two cases is inessential for the discussion here.

For a linear or quasilinear PDE, the characteristic curves are given parametrically by

$$(x_1,\ldots,x_n,u) = (x_1(s),\ldots,x_n(s),u(s))$$

such that the following system of ODEs is satisfied

$$\frac{dx_i}{ds} = a_i(x_1,\ldots,x_n,u) \qquad (2)$$

$$\frac{du}{ds} = c(x_1,\ldots,x_n,u). \qquad (3)$$

Equations (2) and (3) give the characteristics of the PDE.

Fully Nonlinear Case

Consider the partial differential equation

$$F(x_1,\ldots,x_n,u,p_1,\ldots,p_n) = 0 \qquad (4)$$

where the variables p_i are shorthand for the partial derivatives

$$p_i = \frac{\partial u}{\partial x_i}.$$

Let $(x_i(s), u(s), p_i(s))$ be a curve in $\mathbf{R}^{2n+1}$. Suppose that u is any solution, and that

$$u(s) = u(x_1(s), \ldots, x_n(s)).$$

Along a solution, differentiating (4) with respect to s gives

$$\sum_i (F_{x_i} + F_u p_i)\dot{x}_i + \sum_i F_{p_i} \dot{p}_i = 0$$

$$\dot{u} - \sum_i p_i \dot{x}_i = 0$$

$$\sum_i (\dot{x}_i dp_i - \dot{p}_i dx_i) = 0.$$

The second equation follows from applying the chain rule to a solution u, and the third follows by taking an exterior derivative of the relation $du - \sum_i p_i dx_i = 0$. Manipulating these equations gives

$$\dot{x}_i = \lambda F_{p_i}, \quad \dot{p}_i = -\lambda(F_{x_i} + F_u p_i), \quad \dot{u} = \lambda \sum_i p_i F_{p_i}$$

where λ is a constant. Writing these equations more symmetrically, one obtains the Lagrange-Charpit equations for the characteristic

$$\frac{\dot{x}_i}{F_{p_i}} = -\frac{\dot{p}_i}{F_{x_i} + F_u p_i} = \frac{\dot{u}}{\sum p_i F_{p_i}}.$$

Geometrically, the method of characteristics in the fully nonlinear case can be interpreted as requiring that the Monge cone of the differential equation should everywhere be tangent to the graph of the solution.

Example

As an example, consider the advection equation (this example assumes familiarity with PDE notation, and solutions to basic ODEs).

$$a\frac{\partial u}{\partial x} + \frac{\partial u}{\partial t} = 0$$

where a is constant and u is a function of x and t. We want to transform this linear first-order PDE into an ODE along the appropriate curve; i.e. something of the form

$$\frac{d}{ds}u(x(s),t(s)) = F(u,x(s),t(s)),$$

where $(x(s),t(s))$ is a characteristic line. First, we find

$$\frac{d}{ds}u(x(s),t(s)) = \frac{\partial u}{\partial x}\frac{dx}{ds}+\frac{\partial u}{\partial t}\frac{dt}{ds}$$

by the chain rule. Now, if we set $\frac{dx}{ds}=a$ and $\frac{dt}{ds}=1$ we get

$$a\frac{\partial u}{\partial x}+\frac{\partial u}{\partial t}$$

which is the left hand side of the PDE we started with. Thus

$$\frac{d}{ds}u = a\frac{\partial u}{\partial x}+\frac{\partial u}{\partial t}=0.$$

So, along the characteristic line $(x(s),t(s))$, the original PDE becomes the ODE $u_s = F(u,x(s),t(s)) = 0$. That is to say that along the characteristics, the solution is constant. Thus, $u(x_s,t_s) = u(x_0,0)$ where (x_s,t_s) and $(x_0,0)$ lie on the same characteristic. So to determine the general solution, it is enough to find the characteristics by solving the characteristic system of ODEs:

- $\frac{dt}{ds}=1$, letting $t(0)=0$ we know $t=s$,

- $\frac{dx}{ds}=a$, letting $x(0)=x_0$ we know $x=as+x_0=at+x_0$,

- $\frac{du}{ds}=0$, letting $u(0)=f(x_0)$ we know $u(x(t),t)=f(x_0)=f(x-at)$.

In this case, the characteristic lines are straight lines with slope a, and the value of u remains constant along any characteristic line.

Characteristics of Linear Differential Operators

Let X be a differentiable manifold and P a linear differential operator

$$P:C^\infty(X)\to C^\infty(X)$$

of order k. In a local coordinate system x^i,

$$P=\sum_{|\alpha|\le k}P^\alpha(x)\frac{\partial}{\partial x^\alpha}$$

in which α denotes a multi-index. The principal symbol of P, denoted σ_P, is the function on the cotangent bundle T*X defined in these local coordinates by

$$\sigma_P(x,\xi)=\sum_{|\alpha|=k}P^\alpha(x)\xi_\alpha$$

where the ξ_i are the fiber coordinates on the cotangent bundle induced by the coordinate differentials dx^i. Although this is defined using a particular coordinate system, the transformation law relating the ξ_i and the x^i ensures that σ_p is a well-defined function on the cotangent bundle.

The function σ_p is homogeneous of degree k in the ξ variable. The zeros of σ_p, away from the zero section of T^*X, are the characteristics of P. A hypersurface of X defined by the equation $F(x) = c$ is called a characteristic hypersurface at x if

$$\sigma_P(x, dF(x)) = 0.$$

Invariantly, a characteristic hypersurface is a hypersurface whose conormal bundle is in the characteristic set of P.

Qualitative Analysis of Characteristics

Characteristics are also a powerful tool for gaining qualitative insight into a PDE.

One can use the crossings of the characteristics to find shock waves for potential flow in a compressible fluid. Intuitively, we can think of each characteristic line implying a solution to u along itself. Thus, when two characteristics cross, the function becomes multi-valued resulting in a non-physical solution. Physically, this contradiction is removed by the formation of a shock wave, a tangential discontinuity or a weak discontinuity and can result in non-potential flow, violating the initial assumptions.

Characteristics may fail to cover part of the domain of the PDE. This is called a rarefaction, and indicates the solution typically exists only in a weak, i.e. integral equation, sense.

The direction of the characteristic lines indicate the flow of values through the solution, as the example above demonstrates. This kind of knowledge is useful when solving PDEs numerically as it can indicate which finite difference scheme is best for the problem.

Quasilinear First-Order Partial Differential Equations

A first order quasilinear PDE is of the form

$$a(x, y, z)\frac{\partial z}{\partial x} + b(x, y, z)\frac{\partial z}{\partial y} = c(x, y, z). \qquad (1)$$

Such equations occur in a variety of nonlinear wave propagation problems. Let us assume that an integral surface $z = z(x, y)$ of (1) can be found. Writing this integral surface in implicit form as

$$F(x, y, z) = z(x, y) - z = 0$$

Note that the gradient vector $\nabla F = (z_x, z_y, -1)$ is normal to the integral surface $F(x, y, z) = 0$. The equation (1) may be written as

$$az_x + bz_y - c = (a, b, c) \cdot (z_x, z_y, -1) = 0. \qquad (2)$$

This shows that the vector (a, b, c) and the gradient vector ∇F are orthogonal. In other words, the vector (a, b, c) lies in the tangent plane of the integral surface z = z(x, y) at each point in the (x, y, z)-space where $\nabla F \neq 0$.

At each point (x, y, z), the vector (a, b, c) determines a direction in (x, y, z)-space is called the characteristic direction. We can construct a family of curves that have the characteristic direction at each point. If the parametric form of these curves is

$$x = x(t), \ y = y(t) \ \text{ and } \ z = z(t), \tag{3}$$

then we must have

$$\frac{dx}{dt} = a(x(t), y(t), z(t)), \frac{dy}{dt} = b(x(t), y(t), z(t)), \frac{dz}{dt} = c(x(t), y(t), z(t)), \tag{4}$$

because (dx/dt, dy/dt, dz/dt) is the tangent vector along the curves. The solutions of (4) are called the characteristic curves of the quasilinear equation (1).

We assume that a(x, y, z), b(x, y, z), and c(x, y, z) are sufficiently smooth and do not all vanish at the same point. Then, the theory of ordinary diffierential equations ensures that a unique characteristic curve passes through each point (x_o, y_o, z_o). The IVP for(1) requires that z(x, y) be specified on a given curve in (x, y)-space which determines a curve C in (x, y, z)-space referred to as the initial curve. To solve this IVP, we pass a characteristic curve through each point of the initial curve C. If these curves generate a surface known as integral surface. This integral surface is the solution of the IVP.

REMARK. (i) The characteristic equations (4) for x and y are not, in general, uncoupled from the equation for z and hence differ from those in the linear case.

(ii) The characteristics equations (4) can be expressed in the nonparametric form as

$$\frac{dx}{a} = \frac{dy}{b} = \frac{dz}{c}. \tag{5}$$

Below, we shall describe a method for finding the general solution of (1). This method is due to Lagrange hence it is usually referred to as the method of characteristics or the method of Lagrange.

1. The Method of Characteristics

It is a method of solution of quasi-linear PDE which is stated in the following result.

THEOREM. The general solution of the quasi-linear PDE (1) is

$$F(u, v) = 0, \tag{6}$$

where F is an arbitrary function and $u(x, y, z) = c_1$ and $v(x, y, z) = c_2$ form a solution of the equations

$$\frac{dx}{a} = \frac{dy}{b} = \frac{dz}{c}. \tag{7}$$

Proof: If $u(x, y, z) = c_1$ and $v(x, y, z) = c_2$ satisfy the equations (1) then the equations

$$u_x dx + u_y dy + u_z dz = 0$$
$$v_x dx + v_y dy + v_z dz = 0$$

are compatible with (7). Thus, we must have

$$au_x + bu_y + cu_z = 0$$
$$av_x + bv_y + cv_z = 0$$

Solving these equations for a, b and c, we obtain

$$\frac{a}{\dfrac{\partial(u,v)}{\partial(y,z)}} = \frac{b}{\dfrac{\partial(u,v)}{\partial(z,x)}} = \frac{c}{\dfrac{\partial(u,v)}{\partial(x,y)}}. \qquad (8)$$

Differentiate $F(u, v) = 0$ with respect to x and y, respectively, to have

$$\frac{\partial F}{\partial u}\left\{ \frac{\partial u}{\partial x} + \frac{\partial u}{\partial z}\frac{\partial z}{\partial x} \right\} + \frac{\partial F}{\partial v}\left\{ \frac{\partial v}{\partial x} + \frac{\partial v}{\partial z}\frac{\partial z}{\partial x} \right\} = 0$$

$$\frac{\partial F}{\partial u}\left\{ \frac{\partial u}{\partial y} + \frac{\partial u}{\partial z}\frac{\partial z}{\partial y} \right\} + \frac{\partial F}{\partial v}\left\{ \frac{\partial v}{\partial y} + \frac{\partial v}{\partial z}\frac{\partial z}{\partial y} \right\} = 0.$$

Eliminating $\dfrac{\partial F}{\partial u}$ and $\dfrac{\partial F}{\partial v}$ from these equations, we obtain

$$\frac{\partial z}{\partial x}\frac{\partial(u,v)}{\partial(y,z)} + \frac{\partial z}{\partial y}\frac{\partial(u,v)}{\partial(z,x)} = \frac{\partial(u,v)}{\partial(x,y)} \qquad (9)$$

In view of (8), the equation (9) yields

$$a\frac{\partial z}{\partial x} + b\frac{\partial z}{\partial y} = c.$$

Thus, we find that $F(u, v) = 0$ is a solution of the equation (1). This completes the proof.

REMARK. All integral surfaces of the equation (1) are generated by the integral curves of the equations (4).

- All surfaces generated by integral curves of the equations (4) are integral surfaces of the equation (1).

EXAMPLE. Find the general integral of $xz_x + yz_y = z$.

Solution: The associated system of equations are

$$\frac{dx}{x} = \frac{dy}{y} = \frac{dz}{z}.$$

From the first two relations we have

$$\frac{dx}{x} = \frac{dy}{y} \Rightarrow \ln x = \ln y + \ln c_1 \Rightarrow \frac{x}{y} = c_1.$$

Similarly,

$$\frac{dz}{z} = \frac{dy}{y} \Rightarrow \frac{z}{y} = c_2.$$

Take $u_1 = \frac{x}{y}$ and $u_2 = \frac{z}{y}$. The general integral is given by

$$F(\frac{x}{y}, \frac{z}{y}) = 0.$$

EXAMPLE. Find the general integral of the equation

$$z(x+y)z_x + z(x-y)z_y = x^2 + y^2.$$

Solution: The characteristic equations are

$$\frac{dx}{z(x+y)} = \frac{dy}{z(x-y)} = \frac{dz}{x^2+y^2}.$$

Each of these ratios is equivalent to

$$\frac{ydx + xdy - zdz}{0} = \frac{xdx - ydy - zdz}{0}.$$

Consequently, we have

$$d\{xy - \frac{z^2}{2}\} = 0 \quad \text{and} \quad d\{\frac{1}{2}(x^2 - y^2 - z^2)\} = 0.$$

Integrating we obtain two integrals

$$2xy - z^2 = c_1 \quad \text{and} \quad x^2 - y^2 - z^2 = c_2,$$

where c_1 and c_2 are arbitrary constants. Thus, the general solution is

$$F(2xy - z^2, x^2 - y^2 - z^2) = 0,$$

where F is an arbitrary function.

Next, we shall discuss a method for solving a Cauchy problem for the first-order quasi- linear PDE (1). The following theorem gives conditions under which a unique solution of the initial value problem for (1) can be obtained.

THEOREM. Let a(x, y, z), b(x, y, z) and c(x, y, z) in (1) have continuous partial derivatives with respect to x, y and z variables. Let the initial curve C be described parametrically as

$$x = x(s), \quad y = y(s), \quad \text{and} \quad z = z(x(s), y(s)).$$

The initial curve C has a continuous tangent vector and

$$J(s) = \frac{dy}{ds} a\big[x(s), y(s), z(s)\big] - \frac{dx}{ds} b\big[x(s), y(s), z(s)\big] \neq 0 \qquad (10)$$

on C. Then, there exists a unique solution z = z(x, y), defined in some neighborhood of theinitial curve C, satisfies (1) and the initial condition z(x(s), y(s)) = z(s).

Proof: The characteristic system (4) with initial conditions at t = 0 given as x = x(s), y = y(s), and z = z(s) has a unique solution of the form

$$x = x(s, t), \quad y = y(s, t), \quad \text{and} \quad z = z(s, t),$$

with continuous derivatives in s and t, and

$$x(s, 0) = x(s), \quad y(s, 0) = y(s), \quad \text{and} \quad z(s, 0) = z(s).$$

This follows from the existence and uniqueness theory for ODEs. The Jacobian of the transformation x = x(s, t), y = y(s, t) at t = 0 is

$$J(s) = J(s, t)|_{t=0} = \begin{vmatrix} \frac{\partial x}{\partial s} & \frac{\partial x}{\partial t} \\ \frac{\partial y}{\partial s} & \frac{\partial y}{\partial t} \end{vmatrix}_{t=0} = \left[\frac{\partial y}{\partial t} a - \frac{\partial x}{\partial t} b \right]_{t=0} \neq 0. \qquad (11)$$

in view of (10). By the continuity assumption, the Jacobian $J \neq 0$ in a neighbourhood of the initial curve. Thus, by the implicit function theorem, we can solve for s and t as functions of x and y near the initial curve. Then

$$z(s, t) = z(s(x, y), t(x, y)) = Z(x, y).$$

a solution of (1), which can be easily seen as

$$c = \frac{dz}{dt} = \frac{\partial z}{\partial x}\frac{dx}{dt} + \frac{\partial z}{\partial y}\frac{dy}{dt}$$

$$= a\frac{\partial z}{\partial x} + b\frac{\partial z}{\partial y},$$

where we have used (4). The uniqueness of the solution follows from the fact that any two integral surfaces that contain the same initial curve must coincide along all the characteristic curves passing through the initial curve. This is a consequence of the uniqueness theorem for the IVP for (4). This completes our proof.

EXAMPLE. Consider the IVP:

$$\frac{\partial z}{\partial y} + z \frac{\partial z}{\partial x} = 0$$

$$z(x,0) = f(x),$$

where $f(x)$ is a given smooth function.

Solution: We solve this problem using the following steps.

Step 1: (Finding characteristic curves)

To solve the IVP, we parameterize the initial curve as

$$x = s, \quad y = 0, \quad z = f(s).$$

The characteristic equations are

$$\frac{dx}{dt} = z, \quad \frac{dy}{dt} = 1, \quad \frac{dz}{dt} = 0.$$

Let the solutions be denoted as $x(s, t)$, $t(s, t)$, and $z(s, t)$. We immediately find that

$$x(s,t) = zt + s = f(s)t + s, \quad y(s,t) = t, \quad z(s,t) = f(s).$$

where c_i, $i = 1, 2, 3$ are constants to be determined using IC.

Step 2: (Applying IC) The initial conditions at $s = 0$ are given by

$$x(s,0) = s, \quad y(s,0) = 0, \quad \text{and} \quad z(s,0) = f(s).$$

Using these conditions, we obtain

$$x(s,t) = zt + s, \quad y(s,t) = t, \quad \text{and} \quad z(s,t) = f(s).$$

Step 3: (Writing the parametric form of the solution)

The solutions are thus given by

$$x(s,t) = zt + s = f(s,)t + s, \quad y(s,t) = t \quad z(s,t) = f(s).$$

Step 4: (Expressing z (s, t) in terms of z (x, y)) Applying the condition (10), we find that $J(s) = -1 \neq 0$, along the entire initial curve. We can immediately solve for s(x, y) and t(x, y) to obtain

$$s(x,y) = x - tf(s), \quad t(x,y) = y.$$

Since $t = y$ and $s = x - tf(s) = x - yz$, the solution can also be given in implicit form as

$$z = f(x - yz).$$

EXAMPLE. Solve the following quasi-linear PDE:

$$zz_x + yz_y = x, \qquad (x,y) \in R^2$$

subject to the initial condition

$$z(x,1) = 2x, \qquad x \in R.$$

Solution: Here a(x, y, z) = z, b(x, y, z) = y, c(x, y, z) = x. The characteristics equations are

$$\frac{dx}{dt} = z, \quad x(s,0) = s,$$

$$\frac{dy}{dt} = y, \quad y(s,0) = 1,$$

$$\frac{dz}{dt} = x, \quad z(s,0) = 2s.$$

On solving the above ODEs, we obtain

$$x(s,t) = \frac{s}{2}\left(3e^t - e^{-t}\right), \quad y(s,t) = e^t, \quad z(s,t) = \frac{s}{2}\left(3e^t + e^{-t}\right).$$

Solving for (s, t) in terms of (x, y), we obtain

$$s(x,y) = \frac{2xy}{3y^2 - 1}, \quad t(x,y) = \ln(y),$$

$$z(x,y) = z(s(x,y), t(x,y)) = \frac{(3y^2 + 1)x}{(3y^2 - 1)}.$$

Note that the characteristics variables imply that y must be positive (y = e^t). In fact, the solution z is valid only for $3y^2 - 1 > 0$, i.e., for $y > \frac{1}{\sqrt{3}} > 0$. Observe that the change of variables is valid only where

$$\begin{vmatrix} x_s(s,t) & x_t(s,t) \\ y_s(s,t) & y_t(s,t) \end{vmatrix} \neq 0$$

It is easy to verify that this condition leads to $y \neq 1/\sqrt{3}$.

Nonlinear Partial Differential Equation

In mathematics and physics, a nonlinear partial differential equation is a partial differential equation with nonlinear terms. They describe many different physical systems, ranging from gravitation to fluid dynamics, and have been used in mathematics to solve problems such as the Poincaré conjecture and the Calabi conjecture. They are difficult to study: there are almost no general techniques that work for all such equations, and usually each individual equation has to be studied as a separate problem.

Methods for Studying Nonlinear Partial Differential Equations

Existence and Uniqueness of Solutions

A fundamental question for any PDE is the existence and uniqueness of a solution for given boundary conditions. For nonlinear equations these questions are in general very hard: for example, the hardest part of Yau's solution of the Calabi conjecture was the proof of existence for a Monge–Ampere equation.

Singularities

The basic questions about singularities (their formation, propagation, and removal, and regularity of solutions) are the same as for linear PDE, but as usual much harder to study. In the linear case one can just use spaces of distributions, but nonlinear PDEs are not usually defined on arbitrary distributions, so one replaces spaces of distributions by refinements such as Sobolev spaces.

An example of singularity formation is given by the Ricci flow: Richard S. Hamilton showed that while short time solutions exist, singularities will usually form after a finite time. Grigori Perelman's solution of the Poincaré conjecture depended on a deep study of these singularities, where he showed how to continue the solution past the singularities.

Linear Approximation

The solutions in a neighborhood of a known solution can sometimes be studied by linearizing the PDE around the solution. This corresponds to studying the tangent space of a point of the moduli space of all solutions.

Moduli Space of Solutions

Ideally one would like to describe the (moduli) space of all solutions explicitly, and for some very special PDEs this is possible. (In general this is a hopeless problem: it is unlikely that there is any useful description of all solutions of the Navier–Stokes equation for example, as this would involve describing all possible fluid motions.) If the equation has a very large symmetry group, then one is usually only interested in the moduli space of solutions modulo the symmetry group, and this is sometimes a finite-dimensional compact manifold, possibly with singularities; for example, this happens in the case of the Seiberg–Witten equations. A slightly more complicated case is the self dual Yang–Mills equations, when the moduli space is finite-dimensional but not necessarily compact, though it can often be compactified explicitly. Another case when one can sometimes hope to describe all solutions is the case of completely integrable models, when solutions are sometimes a sort of superposition of solitons; for example, this happens for the Korteweg–de Vries equation.

Exact Solutions

It is often possible to write down some special solutions explicitly in terms of elementary functions (though it is rarely possible to describe all solutions like this). One way of finding such explicit solutions is to reduce the equations to equations of lower dimension, preferably ordinary differential equations, which can often be solved exactly. This can sometimes be done using separation of variables, or by looking for highly symmetric solutions.

Some equations have several different exact solutions.

Numerical Solutions

Numerical solution on a computer is almost the only method that can be used for getting information about arbitrary systems of PDEs. There has been a lot of work done, but a lot of work still remains on solving certain systems numerically, especially for the Navier–Stokes and other equations related to weather prediction.

Lax Pair

If a system of PDEs can be put into Lax pair form

$$\frac{dL}{dt} = LA - AL$$

then it usually has an infinite number of first integrals, which help to study it.

Euler–Lagrange Equations

Systems of PDEs often arise as the Euler–Lagrange equations for a variational problem. Systems of this form can sometimes be solved by finding an extremum of the original variational problem.

Integrable Systems

PDEs that arise from integrable systems are often the easiest to study, and can sometimes be completely solved. A well-known example is the Korteweg–de Vries equation.

Symmetry

Some systems of PDEs have large symmetry groups. For example, the Yang–Mills equations are invariant under an infinite-dimensional gauge group, and many systems of equations (such as the Einstein field equations) are invariant under diffeomorphisms of the underlying manifold. Any such symmetry groups can usually be used to help study the equations; in particular if one solution is known one can trivially generate more by acting with the symmetry group.

Sometimes equations are parabolic or hyperbolic "modulo the action of some group": for example, the Ricci flow equation is not quite parabolic, but is "parabolic modulo the action of the diffeomorphism group", which implies that it has most of the good properties of parabolic equations.

The general nonlinear first-order PDE is written in the form

$$F\left(x,y,z,z_x,z_y\right)=0 \qquad (1)$$

where F is not linear in z_x and z_y. Setting $z_x = p$ and $z_y = q$, rewrite (1) as

The Method of Characteristics for Nonlinear PDEs

$$F\left(x,y,z,p,q\right)=0 \qquad (2)$$

Recall the method of characteristics for solving first-order linear PDE:

$$F\left(x,y,z,p,q\right)=a\left(x,y\right)p+b\left(x,y\right)q+c\left(x,y\right)z-d\left(x,y\right)=0$$

In this method, the PDE becomes an ODEs along the characteristics curves which may be regarded as the solutions of the system

$$x'\left(t\right)=a\left(x\left(t\right),y\left(t\right)\right) \text{ and } y'\left(t\right)=b\left(x\left(t\right),y\left(t\right)\right) \qquad (3)$$

Note that $F_p = a(x, y)$ and $F_q = b(x, y)$. Hence, (3) may be written as

$$x'\left(t\right)=F_p \text{ and } y'\left(t\right)=F_q \qquad (4)$$

For solving first-order nonlinear PDE (1), the relation (4) motivates us to define characteristics curves as solutions of the system

$$x'\left(t\right)=F_p\left(x\left(t\right),y\left(t\right),z\left(t\right),p\left(t\right),q\left(t\right)\right) \text{ and } y'\left(t\right)=F_q\left(x\left(t\right),y\left(t\right),z\left(t\right),p\left(t\right),q\left(t\right)\right) \qquad (5)$$

where $z(t) = z(x(t), y(t))$, $p(t) = z_x(x(t), y(t))$, $q(t) = z_y(x(t), y(t))$. However, unlike the linear case, the right sides of (5) depend not only on x(t) and y(t), but also on z(t), p(t) and q(t). Thus, we can expect a large system of five ODEs for the five unknown x(t), y(t), z(t), p(t) and q(t). For the remaining three equations, notice that

$$
\begin{aligned}
z'(t) &= \frac{d}{dt}\{z(x(t),y(t))\} \\
&= z_x x'(t) + z_y y'(t) \\
&= p(t)x'(t) + q(t)y'(t) \\
&= p(t)F_p(x(t),y(t),z(t),p(t),q(t)) + q(t)F_q(x(t),y(t),z(t),p(t),q(t)). \qquad (6)
\end{aligned}
$$

Along a characteristics p is a function of t. The equation for p'(t) is obtained as follows:

$$
\begin{aligned}
p'(t) &= \frac{d}{dt}\{z_x(x(t),y(t))\} \\
&= z_{xx} x'(t) + z_{xy} y'(t) \\
&= z_{xx} F_p(x(t),y(t),z(t),p(t),q(t)) + z_{xy} F_q(x(t),y(t),z(t),p(t),q(t)). \qquad (7)
\end{aligned}
$$

Using the fact that $z(x, y)$ should solve the PDE (1), we obtain

$$0 = \frac{d}{dx}\left\{F\left(x, y, z(x, y), z_x(x, y), z_y(x, y)\right)\right\}$$
$$= F_x + F_z z_x + F_p z_{xx} + F_q z_{yx}$$

Therefore,

$$p'(t) = z_{xx} F_q + z_{yx} F_q = -\left(F_x + p F_z\right). \tag{8}$$

Similarly,

$$q'(t) = -\left[F_y + q F_z\right] \tag{9}$$

Thus, we have the following system of five ODEs

$$x'(t) = F_p(x(t), y(t), z(t), p(t), q(t))$$
$$y'(t) = F_q(x(t), y(t), z(t), p(t), q(t))$$
$$z'(t) = p(t) F_p(x(t), y(t), z(t), p(t), q(t)) + q(t) F_q(x(t), y(t), z(t), p(t), q(t))$$
$$p'(t) = -\{F_x(x(t), y(t), z(t), p(t), q(t)) + p(t) F_z(x(t), y(t), z(t), p(t), q(t))\}$$
$$q'(t) = -\{F_y(x(t), y(t), z(t), p(t), q(t)) + q(t) F_z(x(t), y(t), z(t), p(t), q(t))\} \tag{10}$$

These equations constitute the characteristics system of the PDE (1) and are known as the characteristics equations associated with PDE (1).

NOTE: If the functions which appear in equations (10) satisfy a Lipschitz condition, there is a unique solution of the equations for each prescribed set of initial values of the variables. Therefore the characteristics strip is uniquely determined by any initial element $(x(t_0), y(t_0), z(t_0), p(t_0), q(t_0))$ at any initial point t_0 of t.

An important result about characteristic strips is given below.

THEOREM. The function $F(x, y, z, p, q)$ is a constant along every characteristics strip of the equation $F(x, y, z, p, q) = 0$.

Proof: Along a characteristic strip, we have

$$\begin{aligned}\frac{d}{dt}\{F(x(t), y(t), z(t), p(t), q(t))\} &= F_x x'(t) + F_y y'(t) + F_z z'(t) + F_p p'(t) + F_q q'(t) \\ &= F_x F_p + F_y F_q + F_z(p F_p + q F_q) - F_p(F_x + p F_z) - F_q(F_y + q F_z) \\ &= 0.\end{aligned}$$

This implies $F(x, y, z, p, q) = k$, a constant along the strip.

Solving Cauchy's Problem for Nonlinear PDEs

The objective of this section to solve PDE

$$F\left(x, y, z, z_x, z_y\right) = 0$$

subject to an appropriate initial condition (i.e., z assume prescribed values on some curve).

Let (f (s), g(s)) traces out a regular curve in the xy-plane as s varies. We regard this curve as being an initial curve. We seek a solution u(x, y) of the following problem (known as Cauchy's problem).

$$F\left(x, y, z, z_x, z_y\right) = 0, \quad u\left(f\left(s\right), g\left(s\right)\right) = G\left(s\right) \qquad (11)$$

where G(s) is a continuously differentiable function. Such a problem may have no solution (e.g., the PDE $z^2 + z^2 + 1 = 0$). However, if a solution exists in some neighborhood of theinitial curve, then such a solution can often be determined using the following steps.

Step 1: Find functions h(s) and k(s) (if possible) such that

$$F(f(s), g(s), G(s), h(s), k(s)) = 0, \quad G'(s) = h(s)f'(s) + k(s)g'(s) \text{ and}$$

$$F_p(f(s), g(s), G(s), h(s), k(s))g'(s) - F_q(f(s), g(s), G(s), h(s), k(s))f'(s) \neq 0. \,(12)$$

Note that if h(s) and k(s) do not exist, then (11) has no solution. If there are several choices for (h(s), k(s)), then a solution of (11) exists for each such choice.

Step 2: For each fixed s, solve the following charateristics system for x(s, t), y(s, t), z(s, t), p(s, t),q(s, t) with the given initial conditions p(s, 0) = h(s), q(s, 0) = k(s), where h(s) and k(s) are the functions found in Step 1.

$$
\begin{aligned}
\frac{d}{dt}x(s,t) &= F_p(x(s,t), y(s,t), z(s,t), p(s,t), q(s,t)) \\
\frac{d}{dt}y(s,t) &= F_q(x(s,t), y(s,t), z(s,t), p(s,t), q(s,t)) \\
\frac{d}{dt}z(s,t) &= p(s,t)F_p(x(s,t), y(s,t), z(s,t), p(s,t), q(s,t)) \\
&\quad + q(s,t)F_q(x(s,t), y(s,t), z(s,t), p(s,t), q(s,t)) \qquad (13) \\
\frac{d}{dt}p(s,t) &= -[F_x(x(s,t), y(s,t), z(s,t), p(s,t), q(s,t)) \\
&\quad + p(s,t)F_z(x(s,t), y(s,t), z(s,t), p(s,t), q(s,t))] \\
\frac{d}{dt}q(s,t) &= -[F_y(x(s,t), y(s,t), z(s,t), p(s,t), q(s,t)) \\
&\quad + q(s,t)F_z(x(s,t), y(s,t), z(s,t), p(s,t), q(s,t))]
\end{aligned}
$$

Step 3: As s and t vary, the point (x, y, z), defined by

$$x = x\left(s, t\right), \; y = y\left(s, t\right), \; z = z\left(s, t\right) \qquad (14)$$

traces out the graph of a solution z of (11) in the xyz-space, in a neighborhood of the curve traced out by (f (s), g(s), G(s)). In some cases, one can use the first two equations in (14) to solve for s and t in terms of x and y (say, s = s(x, y) and t = t(x, y)) to obtain a solution z(x, y) = z(s(x, y), t(x, y)), for (x, y) in a neighborhood of the curve (f (s), g(s)).

To illustrate the above steps, let us consider the following example.

EXAMPLE. Solve the PDE $z_x z_y - z = 0$ subject to the condition z(s, −s) = 1.

Solution: Here, we have

$$F(x, y, z, p, q) = pq - z$$

The characteristics system (13) takes the form

$$\frac{dx}{dt} = F_p = q(t), \quad \frac{dy}{dt} = F_q = p(t), \quad \frac{dz}{dt} = pF_p + qF_q = 2p(t)q(t),$$

$$\frac{dp}{dt} = -[F_x + p(t)F_z] = p(t), \quad \frac{dq}{dt} = -[F_y + q(t)F_z] = q(t).$$

Note that

$$\frac{dp}{dt} = p(t) \Rightarrow p(t) = ce^t \quad \text{and} \quad \frac{dq}{dt} = q(t) \Rightarrow q(t) = de^t$$

where c and d are arbitrary constants. Since we are looking for a characteristics strip (i.e., F (x, y, z, p, q) = 0), we set z(t) = p(t)q(t) = cde^{2t}. The equations for the characteristic strip are:

$$x(t) = de^t + d_1, \ y(t) = ce^t + c_1, \ z(t) = cde^{2t}, \ p(t) = ce^t, \ q(t) = de^t$$

where c_1 and d_1 are constants.

The initial condition z(s, −s) = 1 is given on the line y = −x traced out by (s, −s), in (11), we have f (s) = s and g(s) = −s. We must find h(s) and k(s) such that

$$1 = G(s) = h(s)k(s) \quad 0 = G'(s) = h(s) - k(s),$$

$$0 \neq F_p(\ldots)(-1) - F_q(\ldots)(1) = -k(s) - h(s).$$

Thus, we have two choices h(s) = 1 and k(s) = 1, or h(s) = −1 and k(s) = −1. For the choice h(s) = 1 and k(s) = 1, we obtain

$$x(s, t) = e^t - 1 + s, \ y(s, t) = e^t - 1 - s, \ z(s, t) = e^{2t}, \ p(s, t) = e^t, \ q(s, t) = e^t$$

From the first two equations, we obtain

$$e^t = (x + y + z)/2$$

Then the solution is

$$z(x,y) = e^{2t} = \frac{(x+y+2)^2}{4}$$

If we choose h(s) = −1 and k(s) = −1, the solution is given by

$$z(x,y) = \frac{(x+y-2)^2}{4}$$

Compatible Systems and Charpit's Method

In this, we shall study compatible systems of first-order PDEs and the Charpit's method for solving nonlinear PDEs. Let's begin with the following definition.

DEFINITION. (Compatible systems of first-order PDEs) A system of two first-order PDEs

$$F(x,y,z,p,q) = 0 \qquad (1)$$

and

$$g(x,y,z,p,q) = 0 \qquad (2)$$

are said to be compatible if they have a common solution.

THEOREM. The equations f(x, y, z, p, q) = 0 and g(x, y, z, p, q) = 0 are compatible on a domain D if

$$(i)\ J = \frac{\partial(f,g)}{\partial(p,q)} = \begin{vmatrix} f_p & f_q \\ g_p & g_q \end{vmatrix} \neq 0 \, on \, D.$$

(ii)p and q can be explicitly solved from (1) and (2) as p = ϕ(x, y, z) and q = ψ(x, y, z). Further, the equation

$$dz = \phi(x,y,z)dx + \psi(x,y,z)dy$$

is integrable.

THEOREM. A necessary and sufficient condition for the integrability of the equation dz = ϕ(x, y, z)dx + ψ(x, y, z)dy is

$$[f,g] \equiv \frac{\partial(f,g)}{\partial(x,p)} + \frac{\partial(f,g)}{\partial(y,q)} + p\frac{\partial(f,g)}{\partial(z,p)} + q\frac{\partial(f,g)}{\partial(z,q)} = 0 \qquad (3)$$

In other words, the equations (1) and (2) are compatible iff (3) holds.

EXAMPLE. Show that the equations

$$xp - yq = 0, \quad z(xp + yq) = 2xy$$

are compatible and solve them.

Solution: Take $f \equiv xp - yq = 0$, $g \equiv z(xp + yq) - 2xy = 0$. Note that

$$f_x = p, \quad f_y = -q, \quad f_z = 0, \quad f_p = x, \quad f_q = -y$$

and

$$g_x = zp - 2y, \quad g_y = zq - 2x, \quad g_z = xp + yq, \quad g_p = zx, \quad g_q = zy$$

Compute

$$J \equiv \frac{\partial(f,g)}{\partial(p,q)} = \begin{vmatrix} f_p & f_q \\ g_p & g_q \end{vmatrix} = \begin{vmatrix} x & -y \\ zx & zy \end{vmatrix} = zxy + zxy = 2zxy \neq 0$$

for $x \neq 0, y \neq 0, z \neq 0$. Further,

$$\frac{\partial(f,g)}{\partial(x,p)} = \begin{vmatrix} f_x & f_p \\ g_x & g_p \end{vmatrix} = \begin{vmatrix} p & x \\ zp - 2y & zx \end{vmatrix} = zxp - x(zp - 2y) = 2xy$$

$$\frac{\partial(f,g)}{\partial(z,p)} = \begin{vmatrix} f_z & f_p \\ g_z & g_p \end{vmatrix} = \begin{vmatrix} 0 & x \\ xp + yq & zx \end{vmatrix} = 0 - x(xp + yq) = -x^2 p - xyq$$

$$\frac{\partial(f,g)}{\partial(y,q)} = \begin{vmatrix} f_y & f_q \\ g_y & g_q \end{vmatrix} = \begin{vmatrix} -q & -y \\ zq - 2x & zy \end{vmatrix} = -qzy + y(zq - 2x) = -2xy$$

$$\frac{\partial(f,g)}{\partial(z,q)} = \begin{vmatrix} f_z & f_q \\ g_z & g_q \end{vmatrix} = \begin{vmatrix} 0 & -y \\ xp + yq & zy \end{vmatrix} = y(xp + yq) = y^2 q + xyp.$$

It is an easy exercise to verify that

$$\begin{aligned} [f, g] &\equiv \frac{\partial(f,g)}{\partial(x,p)} + \frac{\partial(f,g)}{\partial(y,q)} + p\frac{\partial(f,g)}{\partial(z,p)} + q\frac{\partial(f,g)}{\partial(z,q)} \\ &= 2xy - x^2 p^2 - xypq - 2xy + y^2 q^2 + xypq \\ &= y^2 q^2 - x^2 p^2 \\ &= 0. \end{aligned}$$

So the equations are compatible.

Next step to determine p and q from the two equations $xp-yq = 0$, $z(xp + yq) = 2xy$. Using these two equations, we have

$$zxp + zyq - 2xy = 0 \Rightarrow xp + yq = \frac{2xy}{z}$$

$$\Rightarrow 2xp = \frac{2xy}{z} \Rightarrow p = \frac{y}{z} = \phi(x, y, z)$$

and

$$xp - yq = 0 \Rightarrow q = \frac{xp}{y} = \frac{xy}{yz} = \frac{x}{z}$$

$$\Rightarrow q = \frac{x}{z} = \psi(x, y, z)$$

Substituting p and q in dz = pdx + qdy, we get

$$zdz = ydx + xdy = d(xy)$$

and hence integrating, we obtain

$$z^2 = 2xy + k$$

where k is a constant.

NOTE: For the compatibility of f (x, y, z, p, q) = 0 and g(x, y, z, p, q) = 0 it is not necessary that every solution of f (x, y, z, p, q) = 0 be a solution of g(x, y, z, p, q) = 0 or vice-versa as is generally believed. For instance, the equations

$$f \equiv xp - yq - x = 0 \tag{4}$$

$$g \equiv x^2 p + q - xz = 0 \tag{5}$$

are compatible. They have common solutions z = x + c(1 + xy), where c is an arbitrary constant. Note that z = x(y + 1) is a solution of (4) but not of (5).

Charpit's Method:It is a general method for finding the complete integral of a nonlinear PDE of first-order of the form

$$F(x, y, z, p, q) = 0 \tag{6}$$

Basic Idea: The basic idea of this method is to introduce another partial differential equation of the first order

$$g(x, y, z, p, q, a) = 0 \tag{7}$$

which contains an arbitrary constant a and is such that

(i) Equations (6) and (7) can be solved for p and q to obtain

$$p = p(x, y, z, a), \quad q = q(x, y, z, a)$$

(ii) The equation

$$dz = p(x, y, z, a)\,dx + q(x, y, z, a)\,dy \qquad (8)$$

is integrable.

When such a function g is found, the solution

$$F(x, y, z, a, b) = 0$$

of (8) containing two arbitrary constants a, b will be the solution of (6).

Note: Notice that another PDE g is introduced so that the equations f and g are compatible and then common solutions of f and g are determined in the Charpit's method.

The equations (6) and (7) are compatible if

$$[f, g] \equiv \frac{\partial(f, g)}{\partial(x, p)} + \frac{\partial(f, g)}{\partial(y, q)} + p\frac{\partial(f, g)}{\partial(z, p)} + q\frac{\partial(f, g)}{\partial(z, q)} = 0$$

Expanding it, we are led to the linear PDE

$$f_p \frac{\partial g}{\partial x} + f_q \frac{\partial g}{\partial y} + \left(pf_p + qf_q\right)\frac{\partial g}{\partial z} - \left(f_x + pf_z\right)\frac{\partial g}{\partial p} - \left(f_y + qf_z\right)\frac{\partial g}{\partial q} = 0 \qquad (9)$$

Now solve (9) to determine g by finding the integrals of the following auxiliary equations:

$$\frac{dx}{f_p} = \frac{dy}{f_q} = \frac{dz}{pf_p + qf_q} = \frac{dp}{-\left(f_x + pf_z\right)} = \frac{dq}{-\left(f_y + qf_z\right)} \qquad (10)$$

These equations are known as Charpit's equations which are equivalent to the characteristics equations (10).

Once an integral g(x, y, z, p, q, a) of this kind has been found, the problem reduces to solving for p and q, and then integrating equation (8).

REMARK. For finding integrals, all of Charpit's equations (10) need not to be used.

2. p or q must occur in the solution obtained from (10).

EXAMPLE. Find a complete integral of

$$p^2 x + q^2 y = z \qquad (11)$$

Solution: To find a complete integral, we proceed as follows.

Step 1: (Computing f_x, f_y, f_z, f_p, f_q).

Set $f \equiv p^2x + q^2y - z = 0$. Then

$$f_x = p^2, \quad f_y = q^2, \quad f_z = -1, \quad f_p = 2px, \quad f_q = 2qy$$

$$\Rightarrow pf_p + qf_q = 2p^2x + 2q^2y, \quad -(f_x + pf_z) = -p^2 + p, \quad -(f_y + qf_z) = -q^2 + q$$

Step 2: (Writing Charpit's equations and finding a solution g(x, y, z, p, q, a)).

The Charpit's equations (or auxiliary) equations are:

$$\frac{dx}{f_p} = \frac{dy}{f_q} = \frac{dz}{pf_p + qf_q} = \frac{dp}{-(f_x + pf_z)} = \frac{dq}{-(f_y + qf_z)}$$

$$\Rightarrow \frac{dx}{2px} = \frac{dy}{2qy} = \frac{dz}{2(p^2x + q^2y)} = \frac{dp}{-p^2 + p} = \frac{dq}{-q^2 + q}$$

From which it follows that

$$\frac{p^2dx + 2pxdp}{2p^3x + 2p^2x - 2p^3x} = \frac{q^2dy + 2qydq}{2q^3y + 2q^2y - 2q^3y}$$

$$\Rightarrow \frac{p^2dx + 2pxdp}{p^2x} = \frac{q^2dy + 2qydq}{q^2y}$$

On integrating, we obtain

$$\log(p^2x) = \log(q^2y) + \log a$$

$$\Rightarrow p^2x = aq^2y \qquad\qquad (12)$$

where a is an arbitrary constant.

Step 3: (Solving for p and q).

Using (11) and (12), we find that

$$p^2x + q^2y = z, \quad p^2x = aq^2y$$

$$\implies \quad (aq^2y) + q^2y = z \implies q^2y(1 + a) = z$$

$$\implies \quad q^2 = \frac{z}{(1+a)y} \implies q = \left[\frac{z}{(1+a)y}\right]^{1/2}.$$

and

$$p^2 = aq^2\frac{y}{x} = a\frac{z}{(1+a)y}\frac{y}{x} = \frac{az}{(1+a)x}$$

$$\implies \quad p = \left[\frac{az}{(1+a)x}\right]^{1/2}.$$

Step 4: (Writing dz = p(x, y, z, a)dx + q(x, y, z, a)dy and finding its solution).

Writing

$$dz = \left[\frac{az}{(1+a)x}\right]^{1/2} dx + \left[\frac{z}{(1+a)y}\right]^{1/2} dy$$

$$\Rightarrow \left(\frac{1+a}{z}\right)^{1/2} dz = \left(\frac{a}{x}\right)^{1/2} dx + \left(\frac{1}{y}\right)^{1/2} dy$$

Integrate to have

$$\left[(1+a)z\right]^{1/2} = (ax)^{1/2} + (y)^{1/2} + b$$

the complete integral of the equation (11).

Special Types of First-Order Partial Differential Equations

We shall consider some special types of first-order partial differential equations whose solutions may be obtained easily by Charpit's method.

Type (a): (Equations involving only p and q) If the equation is of the form

$$f(p,q) = 0 \qquad (1)$$

then Charpit's equations take the form

$$\frac{dx}{f_p} = \frac{dy}{f_q} = \frac{dz}{pf_p + qf_q} = \frac{dp}{0} = \frac{dq}{0}$$

An immediate solution is given by p = a, where a is an arbitrary constant. Substituting p = a in (1), we obtain a relation

$$q = Q(a)$$

Then, integrating the expression

$$dz = adx + Q(a)dy$$

we obtain

$$z = ax + Q(a)y + b \qquad (2)$$

where b is a constant. Thus, (2) is a complete integral of (1).

Note: Instead of taking dp = 0, we can take dq = 0 ⇒ q = a. In some problems, taking dq = 0 the amount of computation involved may be reduced considerably.

EXAMPLE. Find a complete integral of the equation pq = 1.

Solution: If $p = a$ then $pq = 1 \Rightarrow q = 1/a$. In this case, $Q(a) = 1/a$. From (2), we obtain a complete integral as

$$z = ax + \frac{y}{a} + b$$

$$\Rightarrow a^2x + y - az = c$$

where a and c are arbitrary constants.

Type (b): (Equations not involving the independent variables) For the equation of the type

$$f(z,p,q) = 0 \qquad (3)$$

Charpit's equation becomes

$$\frac{dx}{f_p} = \frac{dy}{f_q} = \frac{dz}{pf_p + qf_q} = \frac{dp}{-pf_z} = \frac{dq}{-qf_z}$$

From the last two relations, we have

$$\frac{dp}{-pf_z} = \frac{dq}{-qf_z} \Rightarrow \frac{dp}{p} = \frac{dq}{q}$$

$$\Rightarrow p = aq \qquad (4)$$

where a is an arbitrary constant. Solving (3) and (4) for p and q, we obtain

$$q = Q(a,z) \Rightarrow p = aQ(a,z)$$

Now

$$dz = pdx + qdy$$
$$\Rightarrow \quad dz = aQ(a,z)dx + Q(a,z)dy$$
$$\Rightarrow \quad dz = Q(a,z)[adx + dy].$$

It gives complete integral as

$$\int \frac{dz}{Q(a,z)} = ax + y + b \qquad (5)$$

where b is an arbitrary constant.

EXAMPLE. Find a complete integral of the PDE $p^2z^2 + q^2 = 1$.

Solution: Putting $p = aq$ in the given PDE, we obtain

$$a^2q^2z^2 + q^2 = 1$$
$$\Rightarrow \quad q^2\left(1 + a^2z^2\right) = 1$$
$$\Rightarrow \quad q = \left(1 + a^2z^2\right)^{-1/2}$$

Now,

$$p^2 = \left(1 - q^2\right)/z^2 = \left(1 - \frac{1}{\left(1 + a^2z^2\right)}\right)\left(\frac{1}{z^2}\right)$$

$$\Rightarrow p^2 = \frac{a^2}{1 + a^2z^2}$$

$$\Rightarrow p = a\left(1 + a^2z^2\right)^{-1/2}$$

Substituting p and q in dz = pdx + qdy, we obtain

$$dz = a(1 + a^2z^2)^{-1/2}dx + (1 + a^2z^2)^{-1/2}dy$$
$$\Longrightarrow \quad (1 + a^2z^2)^{1/2}dz = adx + dy$$
$$\Longrightarrow \quad \frac{1}{2a}\left\{az(1 + a^2z^2)^{1/2} - \log[az + (1 + a^2z^2)^{1/2}]\right\} = ax + y + b,$$

which is the complete integral of the given PDE.

Type (c): (Separable equations)

A first-order PDE is separable if it can be written in the form

$$f(x,p) = g(y,q) \qquad (6)$$

That is, a PDE in which z is absent and the terms containing x and p can be separated from those containing y and q. For this type of equation, Charpit's equations become

$$\frac{dx}{f_p} = \frac{dy}{-g_q} = \frac{dz}{pf_p + qg_q} = \frac{dp}{-f_x} = \frac{dq}{-g_y}$$

From the last two relation, we obtain an ODE

$$\frac{dp}{-f_x} = \frac{dx}{f_p} \Rightarrow \frac{dp}{dx} + \frac{f_x}{f_p} = 0 \qquad (7)$$

which may be solved to yield p as a function of x and an arbitrary constant a. Writing (7) in the form $f_p dp + f_z dx = 0$, we see that its solution is $f(x, p) = a$. Similarly, we get $g(y, q) = a$. Determine p and q from the equation

$$f(x,p) = a, \quad g(y,q) = a$$

and then use the relation $dz = pdx + qdy$ to determine a complete integral.

EXAMPLE. Find a complete integral of $p^2 y(1 + x^2) = qx^2$.

Solution: First we write the given PDE in the form

$$\frac{p^2(1+x^2)}{x^2} = \frac{q}{y} \quad \text{(separable equation)}$$

It follows that

$$\frac{p^2(1+x^2)}{x^2} = a^2 \Rightarrow p = \frac{ax}{\sqrt{1+x^2}}$$

where a is an arbitrary constant. Similarly,

$$\frac{q}{y} = a^2 \Rightarrow q = a^2 y$$

Now, the relation $dz = pdx + qdy$ yields

$$dz = \frac{ax}{\sqrt{1+x^2}}dx + a^2 ydy \Rightarrow z = a\sqrt{1+x^2} + \frac{a^2 y^2}{2} + b$$

where a and b are arbitrary constant, a complete integral for the given PDE.

Type (d): (Clairaut's equation)

A first-order PDE is said to be in Clairaut form if it can be written as

$$z = px + qy + f(p,q) \qquad (8)$$

Charpit's equations take the form

$$\frac{dx}{x+f_p} = \frac{dy}{y+f_q} = \frac{dz}{px+qy+pf_p+qf_q} = \frac{dp}{0} = \frac{dq}{0}$$

Now, $dp = 0 \Rightarrow p = a$, where a is an arbitrary constant.

$dq = 0 \Rightarrow q = b$, where b is an arbitrary constant.

Substituting the values of p and q in (8), we obtain the required complete integral

$$z = ax + by + f(a, b)$$

EXAMPLE. Find a complete integral of $(p + q)(z - xp - yq) = 1$.

Solution: The given PDE can be put in the form

$$z = xp + yq + \frac{1}{p + q} \qquad (9)$$

which is of Clairaut's type. Putting $p = a$ and $q = b$ in (9), a complete integral is given by

$$z = ax + by + \frac{1}{a + b}$$

where a and b are arbitrary constants.

Jacobi Method

In numerical linear algebra, the Jacobi method (or Jacobi iterative method) is an algorithm for determining the solutions of a diagonally dominantsystem of linear equations. Each diagonal element is solved for, and an approximate value is plugged in. The process is then iterated until it converges. This algorithm is a stripped-down version of the Jacobi transformation method of matrix diagonalization. The method is named after Carl Gustav Jacob Jacobi.

Description

Let

$$A\mathbf{x} = \mathbf{b}$$

be a square system of n linear equations, where:

$$A = \begin{bmatrix} a_{11} & a_{12} & \cdots & a_{1n} \\ a_{21} & a_{22} & \cdots & a_{2n} \\ \vdots & \vdots & \ddots & \vdots \\ a_{n1} & a_{n2} & \cdots & a_{nn} \end{bmatrix}, \qquad \mathbf{x} = \begin{bmatrix} x_1 \\ x_2 \\ \vdots \\ x_n \end{bmatrix}, \qquad \mathbf{b} = \begin{bmatrix} b_1 \\ b_2 \\ \vdots \\ b_n \end{bmatrix}.$$

Then A can be decomposed into a diagonal component D, and the remainder R:

$$A = D + R \qquad \text{where} \qquad D = \begin{bmatrix} a_{11} & 0 & \cdots & 0 \\ 0 & a_{22} & \cdots & 0 \\ \vdots & \vdots & \ddots & \vdots \\ 0 & 0 & \cdots & a_{nn} \end{bmatrix} \text{ and } R = \begin{bmatrix} 0 & a_{12} & \cdots & a_{1n} \\ a_{21} & 0 & \cdots & a_{2n} \\ \vdots & \vdots & \ddots & \vdots \\ a_{n1} & a_{n2} & \cdots & 0 \end{bmatrix}.$$

The solution is then obtained iteratively via

$$\mathbf{x}^{(k+1)} = D^{-1}(\mathbf{b} - R\mathbf{x}^{(k)}),$$

where $\mathbf{x}^{(k)}$ is the kth approximation or iteration of $\mathbf{x}$ and $\mathbf{x}^{(k+1)}$ is the next or $k + 1$ iteration of $\mathbf{x}$. The element-based formula is thus:

$$x_i^{(k+1)} = \frac{1}{a_{ii}}\left(b_i - \sum_{j \neq i} a_{ij} x_j^{(k)} \right), \quad i = 1, 2, \dots, n.$$

The computation of $x_i^{(k+1)}$ requires each element in $\mathbf{x}^{(k)}$ except itself. Unlike the Gauss–Seidel method, we can't overwrite $x_i^{(k)}$ with $x_i^{(k+1)}$, as that value will be needed by the rest of the computation. The minimum amount of storage is two vectors of size n.

Algorithm

Input: initial guess $x^{(0)}$ to the solution, (diagonal dominant) matrix A, right-hand side vector b, convergence criterion

Output: solution when convergence is reached

Comments: pseudocode based on the element-based formula above

while convergence not reached **do**

for i := 1 **step until** n **do**

$\sigma = 0$

for j := 1 **step until** n **do**

if j ≠ i **then**

$\sigma = \sigma + a_{ij} x_j^{(k)}$

end

end

$x_i^{(k+1)} = \frac{1}{a_{ii}}\left(b_i - \sigma \right)$

end

$k = k + 1$

end

Convergence

The standard convergence condition (for any iterative method) is when the spectral radius of the iteration matrix is less than 1:

$$\rho(D^{-1}R) < 1.$$

A sufficient (but not necessary) condition for the method to converge is that the matrix A is strictly or irreducibly diagonally dominant. Strict row diagonal dominance means that for each row, the absolute value of the diagonal term is greater than the sum of absolute values of other terms:

$$|a_{ii}| > \sum_{j \neq i} |a_{ij}|.$$

The Jacobi method sometimes converges even if these conditions are not satisfied.

Example

A linear system of the form $Ax = b$ with initial estimate $x^{(0)}$ is given by

$$A = \begin{bmatrix} 2 & 1 \\ 5 & 7 \end{bmatrix}, b = \begin{bmatrix} 11 \\ 13 \end{bmatrix} \quad \text{and} \quad x^{(0)} = \begin{bmatrix} 1 \\ 1 \end{bmatrix}.$$

We use the equation $x^{(k+1)} = D^{-1}(b - Rx^{(k)})$, described above, to estimate x. First, we rewrite the equation in a more convenient form $D^{-1}(b - Rx^{(k)}) = Tx^{(k)} + C$, where $T = -D^{-1}R$ and $C = D^{-1}b$. Note that $R = L + U$ where L and U are the strictly lower and upper parts of A. From the known values

$$D^{-1} = \begin{bmatrix} 1/2 & 0 \\ 0 & 1/7 \end{bmatrix}, L = \begin{bmatrix} 0 & 0 \\ 5 & 0 \end{bmatrix} \quad \text{and} \quad U = \begin{bmatrix} 0 & 1 \\ 0 & 0 \end{bmatrix}.$$

we determine $T = -D^{-1}(L + U)$ as

$$T = \begin{bmatrix} 1/2 & 0 \\ 0 & 1/7 \end{bmatrix} \left\{ \begin{bmatrix} 0 & 0 \\ -5 & 0 \end{bmatrix} + \begin{bmatrix} 0 & -1 \\ 0 & 0 \end{bmatrix} \right\} = \begin{bmatrix} 0 & -1/2 \\ -5/7 & 0 \end{bmatrix}.$$

Further, C is found as

$$C = \begin{bmatrix} 1/2 & 0 \\ 0 & 1/7 \end{bmatrix} \begin{bmatrix} 11 \\ 13 \end{bmatrix} = \begin{bmatrix} 11/2 \\ 13/7 \end{bmatrix}.$$

With T and C calculated, we estimate x as $x^{(1)} = Tx^{(0)} + C$:

$$x^{(1)} = \begin{bmatrix} 0 & -1/2 \\ -5/7 & 0 \end{bmatrix} \begin{bmatrix} 1 \\ 1 \end{bmatrix} + \begin{bmatrix} 11/2 \\ 13/7 \end{bmatrix} = \begin{bmatrix} 5.0 \\ 8/7 \end{bmatrix} \approx \begin{bmatrix} 5 \\ 1.143 \end{bmatrix}.$$

The next iteration yields

$$x^{(2)} = \begin{bmatrix} 0 & -1/2 \\ -5/7 & 0 \end{bmatrix} \begin{bmatrix} 5.0 \\ 8/7 \end{bmatrix} + \begin{bmatrix} 11/2 \\ 13/7 \end{bmatrix} = \begin{bmatrix} 69/14 \\ -12/7 \end{bmatrix} \approx \begin{bmatrix} 4.929 \\ -1.714 \end{bmatrix}.$$

This process is repeated until convergence (i.e., until $\| Ax^{(n)} - b \|$ is small). The solution after 25 iterations is

$$x = \begin{bmatrix} 7.111 \\ -3.222 \end{bmatrix}.$$

Another example

Suppose we are given the following linear system:

$$10x_1 - x_2 + 2x_3 = 6,$$
$$-x_1 + 11x_2 - x_3 + 3x_4 = 25,$$
$$2x_1 - x_2 + 10x_3 - x_4 = -11,$$
$$3x_2 - x_3 + 8x_4 = 15.$$

If we choose (0, 0, 0, 0) as the initial approximation, then the first approximate solution is given by

$$x_1 \qquad = (6 + 0 - 0)/10 = 0.6,$$
$$x_2 = (25 - 0 - 0)/11 = 25/11 = 2.2727,$$
$$x_3 \qquad = (-11 - 0 - 0)/10 = -1.1,$$
$$x_4 \qquad = (15 - 0 - 0)/8 = 1.875.$$

Using the approximations obtained, the iterative procedure is repeated until the desired accuracy has been reached. The following are the approximated solutions after five iterations.

x_1	x_2	x_3	x_4
0.6	2.27272	-1.1	1.875
1.04727	1.7159	-0.80522	0.88522
0.93263	2.05330	-1.0493	1.13088
1.01519	1.95369	-0.9681	0.97384
0.98899	2.0114	-1.0102	1.02135

The exact solution of the system is (1, 2, −1, 1).

An example using Python and Numpy

The following numerical procedure simply iterates to produce the solution vector.

```
import numpy as np

ITERATION_LIMIT = 1000

# initialize the matrix

A = np.array    ([[10.,-1.,2.,0.],

                 [-1.,11.,-1.,3.],
```

```
                     [2.,-1.,10.,-1.],
                     [0.0,3.,-1.,8.]])
# initialize the RHS vector
b = np.array     ([6.,25.,-11.,15.])
# prints the system
print("System:")
for i in range(A.shape[0]):
    row = ["{}*x{}".format(A[i,j],j+1)for j in range(A.shape[1])]
    print(" + ".join(row),"=",b[i])
print()
x = np.zeros_like(b)
for it_count in range(ITERATION_LIMIT):
    print("Current solution:",x)
    x_new = np.zeros_like(x)
    for i in range(A.shape):
        s1 = np.dot(A[i,:i],x[:i])
        s2 = np.dot(A[i,i+1:],x[i+1:])
        x_new[i]=(b[i]-s1-s2)/A[i,i]
    if np.allclose(x,x_new,atol=1e-10,rtol=0.):
        break
    x = x_new
print("Solution:")
print(x)
error = np.dot(A,x)-b
print("Error:")
print(error)
```

Produces the Output:

```
System:
10.0*x1 + -1.0*x2 + 2.0*x3 + 0.0*x4 = 6.0
-1.0*x1 + 11.0*x2 + -1.0*x3 + 3.0*x4 = 25.0
```

```
2.0*x1 + -1.0*x2 + 10.0*x3 + -1.0*x4 = -11.0
0.0*x1 + 3.0*x2 + -1.0*x3 + 8.0*x4 = 15.0
Current solution: [ 0.   0.   0.   0.]
Current solution: [ 0.6          2.27272727 -1.1          1.875      ]
Current solution: [ 1.04727273  1.71590909 -0.80522727  0.88522727]
Current solution: [ 0.93263636  2.05330579 -1.04934091  1.13088068]
Current solution: [ 1.01519876  1.95369576 -0.96810863  0.97384272]
Current solution: [ 0.9889913   2.01141473 -1.0102859   1.02135051]
Current solution: [ 1.00319865  1.99224126 -0.99452174  0.99443374]
Current solution: [ 0.99812847  2.00230688 -1.00197223  1.00359431]
Current solution: [ 1.00062513  1.9986703  -0.99903558  0.99888839]
Current solution: [ 0.99967415  2.00044767 -1.00036916  1.00061919]
Current solution: [ 1.0001186   1.99976795 -0.99982814  0.99978598]
Current solution: [ 0.99994242  2.00008477 -1.00006833  1.0001085 ]
Current solution: [ 1.00002214  1.99995896 -0.99996916  0.99995967]
Current solution: [ 0.99998973  2.00001582 -1.00001257  1.00001924]
Current solution: [ 1.00000409  1.99999268 -0.99999444  0.9999925 ]
Current solution: [ 0.99999816  2.00000292 -1.0000023   1.00000344]
Current solution: [ 1.00000075  1.99999868 -0.99999899  0.99999862]
Current solution: [ 0.99999967  2.00000054 -1.00000042  1.00000062]
Current solution: [ 1.00000014  1.99999976 -0.99999982  0.99999975]
Current solution: [ 0.99999994  2.0000001  -1.00000008  1.00000011]
Current solution: [ 1.00000003  1.99999996 -0.99999997  0.99999995]
Current solution: [ 0.99999999  2.00000002 -1.00000001  1.00000002]
Current solution: [ 1.          1.99999999 -0.99999999  0.99999999]
Current solution: [ 1.  2. -1.  1.]
Solution:
[ 1.  2. -1.  1.]
Error:
[ -2.81440107e-08   5.15706873e-08  -3.63466359e-08   4.17092547e-08]
```

Weighted Jacobi Method

The weighted Jacobi iteration uses a parameter ω to compute the iteration as

$$x^{(k+1)} = \omega D^{-1}(b - Rx^{(k)}) + (1-\omega)x^{(k)}$$

with $\omega = 2/3$ being the usual choice.

Recent Developments

In 2014, a refinement of the algorithm, called *scheduled relaxation Jacobi (SRJ) method*, was published. The new method employs a schedule of over- and under-relaxations and provides performance improvements for solving elliptic equations discretized on large two- and three-dimensional Cartesian grids. The described algorithm applies the well-known technique of polynomial (Chebyshev) acceleration to a problem with a known spectrum distribution that can be classified either as a multi-step method or a one-step method with a non-diagonal preconditioner. However, none of them are Jacobi-like methods.

Improvements published in 2015.

Consider the following first-order PDE of the form

$$f\left(x, y, z, u_x, u_y, u_z\right) = 0 \qquad (1)$$

where x, y, z are independent variables and u is the dependent variable. Note that the dependent variable u does not appear in the PDE (1).

Idea of Jacobi's Method: The fundamental idea of Jacobi's method is to introduce two first-order PDEs involving two arbitrary constants a and b of the following form

$$h_1\left(x, y, z, u_x, u_y, u_z, a\right) = 0 \qquad (2)$$

$$h_2\left(x, y, z, u_x, u_y, u_z, b\right) = 0 \qquad (3)$$

such that

$$\frac{\partial(f, h_1, h_2)}{\partial\left(u_x, u_y, u_z\right)} \neq 0 \qquad (4)$$

and

- Equations (1), (2) and (3) can be solved for u_x, u_y, u_z;

- The equation

$$du = u_x dx + u_y dy + u_z dz \qquad (5)$$

is integrable.

If $h_1 = 0$ and $h_2 = 0$ are compatible with $f = 0$ then h_1 and h_2 satisfy

$$\frac{\partial(f,h)}{\partial(x,u_x)} + \frac{\partial(f,h)}{\partial(y,u_y)} + \frac{\partial(f,h)}{\partial(z,u_z)} = 0 \qquad (6)$$

for $h = h_i$, $i = 1, 2$. Equation (6) leads to a semi-linear PDE of the form

$$f_{u_x}\frac{\partial h}{\partial x} + f_{u_y}\frac{\partial h}{\partial y} + f_{u_z}\frac{\partial h}{\partial z} - f_x\frac{\partial h}{\partial u_x} - f_y\frac{\partial h}{\partial u_y} - f_z\frac{\partial h}{\partial u_z} \qquad (7)$$

for $h = h_i$, $i = 1, 2$. The associated auxiliary equations are given by

$$\frac{dx}{f_{u_x}} = \frac{dy}{f_{u_y}} = \frac{dz}{f_{u_z}} = \frac{du_x}{-f_x} = \frac{du_y}{-f_y} = \frac{du_z}{-f_z} \qquad (8)$$

The rest of the procedure is same as in Charpit's method.

The method just described above can be applied to solve first-order equation of the form

$$f(x,y,z,p,q) = 0 \qquad (9)$$

Next, we shall show how to transform the equation $f(x, y, z, p, q) = 0$ into the equation $g(x, y, z, u_x, u_y, u_z) = 0$ so that the above procedure can be applied.

If $u(x, y, z)$ is a relation between x, y and z and satisfies (9) then we have

$$u_x + u_z z_x = 0 \Rightarrow u_x + u_z p = 0 \Rightarrow p = -u_x / u_z$$
$$u_y + u_z z_y = 0 \Rightarrow u_y + u_z q = 0 \Rightarrow q = -u_y / u_z$$

Substituting

$$p = -u_x / u_z \qquad \text{and} \qquad q = -u_y / u_z$$

in (9) we obtain an equation

$$g(x,y,z,u_x,u_y,u_z) = 0 \qquad (10)$$

which can be solved by the Jacobi's method.

EXAMPLE. Find a complete integral of $p^2 x + q^2 y = z$ by Jacobi's method.

Step 1: (Converting the given PDE into the form $f(x, y, z, u_x, u_y, u_z) = 0$).

Set $p = -u_x/u_z$ and $q = -u_y/u_z$ in the given PDE to obtain

$$\frac{u_x^2}{u_z^2}x + \frac{u_y^2}{u_z^2}y = z$$

$$\Rightarrow xu_x^2 + yu_y^2 - zu_z^2 = 0$$

Thus

$$f\left(x, y, z, u_x, u_y, u_z\right) = xu_x^2 + yu_y^2 - zu_z^2 = 0 \qquad (11)$$

Step 2: Solving PDE (11) by Jacobi's method

Step 2(a): Compute $f_{u_x}, f_{u_y}, f_{u_z}, f_x, f_y, f_z$

$$f_{u_x} = 2xu_x, \quad f_{u_y} = 2yu_y, \quad f_{u_z} = 2zu_z, \quad f_x = u_x^2, \quad f_y = u_y^2, \quad f_z = -u_z^2$$

Step 2(b): Writing auxiliary equation and solving for u_x, u_y and u_z. The auxiliary equations are given by

$$\frac{dx}{f_{u_x}} = \frac{dy}{f_{u_y}} = \frac{dz}{f_{u_z}} = \frac{du_x}{-f_x} = \frac{du_y}{-f_y} = \frac{du_z}{-f_z}$$

$$\Rightarrow \frac{dx}{2xu_x} = \frac{dy}{2yu_y} = \frac{dz}{-2zu_z} = \frac{du_x}{-u_x^2} = \frac{du_y}{-u_y^2} = \frac{du_z}{u_z^2}$$

Now,

$$\frac{dx}{2xu_x} = \frac{du_x}{-u_x^2} \implies \frac{u_x dx}{2xu_x^2} = \frac{-2x du_x}{2xu_x^2}$$

$$\implies u_x dx = -2x du_x$$

$$\implies \frac{dx}{x} = -2\frac{du_x}{u_x}$$

$$\implies \log x = -2\log(u_x) + \log(a)$$

$$\implies \log x + \log(u_x^2) = \log(a)$$

$$\implies xu_x^2 = a \implies u_x = \left(\frac{a}{x}\right)^{1/2}.$$

Similarly, we get

$$yu_y^2 = b \implies u_y = \left(\frac{b}{y}\right)^{1/2}$$

and

$$u_z = \left[\frac{(a+b)}{z}\right]^{1/2}$$

Step 2(c): Solving the equation $du = u_x dx + u_y dy + u_z dz$.

$$du = \left(\frac{a}{x}\right)^{1/2} dx + \left(\frac{b}{y}\right)^{1/2} dy + \left(\frac{a+b}{z}\right)^{1/2} dz$$

$$\Rightarrow u = 2(ax)^{1/2} + 2(by)^{1/2} + 2\left((a+b)z\right)^{1/2} + c \qquad (12)$$

Step 3: (Finding solution of the given PDE from the solution of PDE (11)).

Writing $u = c$ in (12), we get the complete integral of the given PDE as

$$z = \left[\left(\frac{ax}{a+b}\right)^{1/2} + \left(\frac{by}{a+b}\right)^{1/2}\right]^2$$

References

- Polyanin, A. D.; Zaitsev, V. F.; Moussiaux, A. (2002), Handbook of First Order Partial Differential Equations, London: Taylor & Francis, ISBN 0-415-27267-X

- Sarra, Scott (2003), "The Method of Characteristics with applications to Conservation Laws", Journal of Online Mathematics and its Applications

- Johns Hopkins University (June 30, 2014). "19th century math tactic gets a makeover—and yields answers up to 200 times faster". Phys.org. Douglas, Isle Of Man, United Kingdom: Omicron Technology Limited. Retrieved 2014-07-01

- Roubíček, T. (2013), Nonlinear Partial Differential Equations with Applications (2nd ed.), Basel, Boston, Berlin: Birkhäuser, ISBN 978-3-0348-0512-4, MR MR3014456

- Yang, Xiang; Mittal, Rajat (June 27, 2014). "Acceleration of the Jacobi iterative method by factors exceeding 100 using scheduled relaxation". Journal of Computational Physics. doi:10.1016/j.jcp.2014.06.010

- Delgado, Manuel (1997), "The Lagrange-Charpit Method", SIAM Review, 39 (2): 298–304, Bibcode:1997SIAMR..39..298D, JSTOR 2133111, doi:10.1137/S0036144595293534

- Pokhozhaev, S.I. (2001), "Non-linear partial differential equation", in Hazewinkel, Michiel, Encyclopedia of Mathematics, Springer, ISBN 978-1-55608-010-4

Partial Differential Equations of Second-Order

3

Second order partial differential equation studies problems like heat flow, magnetism, electricity and fluid dynamics. They can be classified into hyperbolic, parabolic or elliptic. The two examples of second-order partial differential equations cited in the chapter are Laplace equation and Poisson equation. The topics discussed in the chapter are of great importance to broaden the existing knowledge on second-order partial differential equations.

Elliptic Partial Differential Equation

Second order linear partial differential equations (PDEs) are classified as either elliptic, hyperbolic, or parabolic. Any second order linear PDE in two variables can be written in the form

$$Au_{xx} + 2Bu_{xy} + Cu_{yy} + Du_x + Eu_y + Fu + G = 0,$$

where A, B, C, D, E, F, and G are functions of x and y and where $u_x = \dfrac{\partial u}{\partial x}$ and similarly for $u_{xx}, u_y, u_{yy}, u_{xy}$. A PDE written in this form is elliptic if

$$B^2 - AC < 0,$$

with this naming convention inspired by the equation for a planar ellipse.

The simplest nontrivial examples of elliptic PDE's are the Laplace equation, $\nabla^2 u = u_{xx} + u_{yy} = 0$, and the Poisson equation, $\nabla^2 u = u_{xx} + u_{yy} = f(x, y)$. In a sense, any other elliptic PDE in two variables can be considered to be a generalization of one of these equations, as it can always be put into the canonical form

$$u_{xx} + u_{yy} + \text{ (lower-order terms)} = 0$$

through a change of variables.

Qualitative Behavior

Elliptic equations have no real characteristic curves, curves along which it is not possible to eliminate at least one second derivative of u from the conditions of the Cauchy problem. Since characteristic curves are the only curves along which solutions to partial differential equations with smooth parameters can have discontinuous derivatives, solutions to elliptic equations cannot have discontinuous derivatives anywhere. This means elliptic equations are well suited to describe equilibrium states, where any discontinuities have already been smoothed out. For instance, we can obtain Laplace's equation from the heat equation $u_t = \nabla^2 u$ by setting $u_t = 0$. This means that Laplace's equation describes a steady state of the heat equation.

In parabolic and hyperbolic equations, characteristics describe lines along which information about the initial data travels. Since elliptic equations have no real characteristic curves, there is no meaningful sense of information propagation for elliptic equations. This makes elliptic equations better suited to describe static, rather than dynamic, processes.

Derivation of Canonical Form

We derive the canonical form for elliptic equations in two variables,
$u_{xx} + u_{yy} +$ (lower-order terms) $= 0$.

$$\xi = \xi(x, y) \text{ and } \eta = \eta(x, y).$$

If $u(\xi, \eta) = u[\xi(x, y), \eta(x, y)]$, applying the chain rule once gives

$$u_x = u_\xi \xi_x + u_\eta \eta_x \text{ and } u_y = u_\xi \xi_y + u_\eta \eta_y,$$

a second application gives

$$u_{xx} = u_{\xi\xi}\xi^2_x + u_{\eta\eta}\eta^2_x + 2u_{\xi\eta}\xi_x\eta_x + u_\xi\xi_{xx} + u_\eta\eta_{xx},$$

$$u_{yy} = u_{\xi\xi}\xi^2_y + u_{\eta\eta}\eta^2_y + 2u_{\xi\eta}\xi_y\eta_y + u_\xi\xi_{yy} + u_\eta\eta_{yy}, \text{ and}$$

$$u_{xy} = u_{\xi\xi}\xi_x\xi_y + u_{\eta\eta}\eta_x\eta_y + u_{\xi\eta}(\xi_x\eta_y + \xi_y\eta_x) + u_\xi\xi_{xy} + u_\eta\eta_{xy}.$$

We can replace our PDE in x and y with an equivalent equation in ξ and η

$$au_{\xi\xi} + 2bu_{\xi\eta} + cu_{\eta\eta} + \text{(lower-order terms)} = 0,$$

where

$$a = A\xi_x^2 + 2B\xi_x\xi_y + C\xi_y^2,$$

$$b = 2A\xi_x\eta_x + 2B(\xi_x\eta_y + \xi_y\eta_x) + 2C\xi_y\eta_y, \text{ and}$$

$$c = A\eta_x^2 + 2B\eta_x\eta_y + C\eta_y^2.$$

To transform our PDE into the desired canonical form, we seek ξ and η such that $a = c$ and $b = 0$. This gives us the system of equations

$$a - c = A(\xi_x^2 - \eta_x^2) + 2B(\xi_x\xi_y - \eta_x\eta_y) + C(\xi_y^2 - \eta_y^2) = 0$$

$$b = 0 = 2A\xi_x\eta_x + 2B(\xi_x\eta_y + \xi_y\eta_x) + 2C\xi_y\eta_y,$$

Adding i times the second equation to the first and setting $\phi = \xi + i\eta$ gives the quadratic equation

$$A\phi_x^2 + 2B\phi_x\phi_y + C\phi_y^2 = 0.$$

Since the discriminant $B^2 - AC < 0$, this equation has two distinct solutions,

$$\phi_x \phi_y \quad \frac{B \pm i\sqrt{AC - B}}{A}$$

which are complex conjugates. Choosing either solution, we can solve for $\phi(x,y)$, and recover ξ and η with the transformations $\xi = \text{Re}\,\phi$ and $\eta = \text{Im}\,\phi$. Since η and ξ will satisfy $a - c = 0$ and $b = 0$, so with a change of variables from x and y to η and ξ will transform the PDE

$$Au_{xx} + 2Bu_{xy} + Cu_{yy} + Du_x + Eu_y + Fu + G = 0,$$

into the canonical form

$$u_{\xi\xi} + u_{\eta\eta} + \text{(lower-order terms)} = 0,$$

as desired.

In Higher Dimensions

A general second order partial differential equation in n variables takes the form

$$\sum_{i=1}^{n}\sum_{j=1}^{n} a_{i,j} \frac{\partial^2 u}{\partial x_i \partial x_j} + \text{(lower-order terms)} = 0.$$

This equation is considered elliptic if there are no characteristic surfaces, i.e. surfaces along which it is not possible to eliminate at least one second derivative of u from the conditions of the Cauchy problem.

Unlike the two dimensional case, this equation cannot in general be reduced to a simple canonical form.

Parabolic Partial Differential Equation

A parabolic partial differential equation is a type of second-order partial differential equation (PDE) of the form

$$Au_{xx} + 2Bu_{xy} + Cu_{yy} + Du_x + Eu_y + F = 0$$

that satisfies the condition

$$B^2 - AC = 0.$$

This definition is analogous to the definition of a planar parabola.

This form of partial differential equation is used to describe a wide family of problems in science including heat diffusion, ocean acoustic propagation, physical or mathematical systems with a time variable, and processes that behave essentially like heat diffusing through a solid.

A simple example of a parabolic PDE is the one-dimensional heat equation,

$$u_t = ku_{xx},$$

where $u(t,x)$ is the temperature at time t and at position x, and k is a constant. The symbol u_t signifies the partial derivative with respect to the time variable t, and similarly u_{xx} is the second partial derivative with respect to x. For this example, t replaced the role of y in the equation determining the type.

This equation says, roughly, that temperature at a given time and point rises or falls at a rate proportional to the difference between the temperature at that point and the average temperature near that point. The quantity u_{xx} measures how far off the temperature is from satisfying the mean value property of harmonic functions.

A generalization of the heat equation is

$$u_t = -Lu,$$

where   is a second-order elliptic operator (implying L must be positive also; a case where L is non-positive is described below). Such a system can be hidden in an equation of the form

$$\nabla \cdot (a(x)\nabla u(x)) + b(x)^T \nabla u(x) + cu(x) = f(x)$$

if the matrix-valued function $a(x)$ has a kernel of dimension 1.

Solution

Under broad assumptions, parabolic PDEs as given above have solutions for all x,y and $t > 0$. An equation of the form $u_t = -L(u)$ is considered parabolic if L is a (possibly nonlinear) function of u and its first and second derivatives, with some further conditions on L. With such a nonlinear parabolic differential equation, solutions exist for a short time—but may explode in a singularity in a finite amount of time. Hence, the difficulty is in determining solutions for all time, or more generally studying the singularities that arise. This is in general quite difficult, as in the solution of the Poincaré conjecture via Ricci flow.

Backward Parabolic Equation

One may occasionally wish to consider PDEs of the form $u_t = L(u)$ where L is a positive elliptic operator. While these problems are no longer necessarily well-posed (solutions may grow unbounded in finite time, or not even exist), they occur when studying the reflection of singularities of solutions to various other PDEs.

This class of equations is closely related to standard hyperbolic equations, which can be seen easily by considering the so-called 'backwards heat equation':

$$\begin{cases} u_t = \Delta u & \text{on } \Omega \times (0,T), \\ u = 0 & \text{on } \partial\Omega \times (0,T), \\ u = f & \text{on } \Omega \times \{T\}. \end{cases}$$

This is essentially the same as the backward hyperbolic equation:

$$\begin{cases} u_t = -\Delta u & \text{on } \Omega \times (0, T), \\ u = 0 & \text{on } \partial\Omega \times (0, T), \\ u = f & \text{on } \Omega \times \{0\}. \end{cases}$$

Examples

- Heat equation
- Mean curvature flow
- Ricci flow

Heat Equation

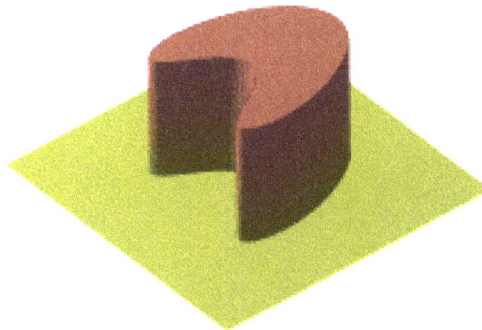

In this example, the heat equation in two dimensions predicts that if one area of an otherwise cool metal plate has been heated, say with a torch, over time the temperature of that area will gradually decrease, starting at the edge and moving inward. Meanwhile the part of the plate outside that region will be getting warmer. Eventually the entire plate will reach a uniform intermediate temperature. Both height and color are used to show temperature.

The heat equation is a parabolic partial differential equation that describes the distribution of heat (or variation in temperature) in a given region over time.

Statement of the Equation

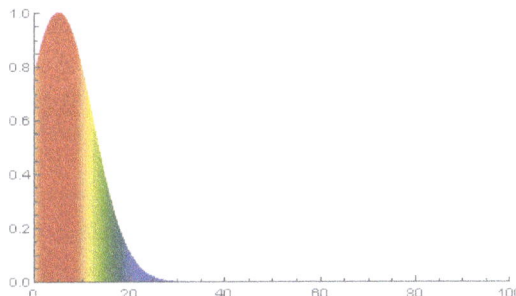

The behaviour of temperature when the sides of a 1D rod are at fixed temperatures (in this case, 0.8 and 0 with initial Gaussian distribution). The temperature becomes linear function, because that is the stable solution of the equation: wherever temperature has a nonzero second spatial derivative, the time derivative is nonzero as well.

For a function $u(x,y,z,t)$ of three spatial variables (x,y,z) and the time variable t, the heat equation is

$$\frac{\partial u}{\partial t} - \alpha \left(\frac{\partial^2 u}{\partial x^2} + \frac{\partial^2 u}{\partial y^2} + \frac{\partial^2 u}{\partial z^2} \right) = 0$$

More generally in any coordinate system:

$$\frac{\partial u}{\partial t} - \alpha \nabla^2 u = 0$$

where α is a positive constant, and Δ or ∇^2 denotes the Laplace operator. In the physical problem of temperature variation, $u(x,y,z,t)$ is the temperature and α is the thermal diffusivity. For the mathematical treatment it is sufficient to consider the case $\alpha = 1$.

Note that the state equation, given by the first law of thermodynamics (i.e. conservation of energy), is written in the following form (assuming no mass transfer or radiation). This form is more general and particularly useful to recognize which property (e.g. c_p or ρ) influences which term.

$$\rho c_p \frac{\partial T}{\partial t} - \nabla \cdot \left(k \nabla T \right) = \dot{q}_V$$

where $\dot{q}_V$ is the volumetric heat flux.

The heat equation is of fundamental importance in diverse scientific fields. In mathematics, it is the prototypical parabolic partial differential equation. In probability theory, the heat equation is connected with the study of Brownian motion via the Fokker–Planck equation. In financial mathematics it is used to solve the Black–Scholes partial differential equation. The diffusion equation, a more general version of the heat equation, arises in connection with the study of chemical diffusion and other related processes.

General Description

Suppose one has a function u that describes the temperature at a given location (x, y, z). This function will change over time as heat spreads throughout space. The heat equation is used to determine the change in the function u over time. The rate of change of u is proportional to the "curvature" of u. Thus, the sharper the corner, the faster it is rounded off. Over time, the tendency is for peaks to be eroded, and valleys filled in. If u is linear in space (or has a constant gradient) at a given point, then u has reached steady-state and is unchanging at this point (assuming a constant thermal conductivity).

One of the interesting properties of the heat equation is the maximum principle that says that the maximum value of u is either earlier in time than the region of concern or on the edge of the region of concern. This is essentially saying that temperature comes either from some source or from earlier in time because heat permeates but is not created from nothingness. This is a property of parabolic partial differential equations and is not difficult to prove mathematically.

Another interesting property is that even if u has a discontinuity at an initial time $t = t_0$, the temperature becomes smooth as soon as $t > t_0$. For example, if a bar of metal has temperature 0 and another has temperature 100 and they are stuck together end to end, then very quickly the temperature at the point of connection will become 50 and the graph of the temperature will run smoothly from 0 to 50.

The heat equation is used in probability and describes random walks. It is also applied in financial mathematics for this reason.

It is also important in Riemannian geometry and thus topology: it was adapted by Richard S. Hamilton when he defined the Ricci flow that was later used by Grigori Perelman to solve the topological Poincaré conjecture.

The Physical Problem and the Equation

Derivation in One Dimension

The heat equation is derived from Fourier's law and conservation of energy (Cannon 1984). By Fourier's law, the rate of flow of heat energy per unit area through a surface is proportional to the negative temperature gradient across the surface,

$$q = -k \nabla u$$

where k is the thermal conductivity and u is the temperature. In one dimension, the gradient is an ordinary spatial derivative, and so Fourier's law is

$$q = -k \frac{\partial u}{\partial x}$$

In the absence of work done, a change in internal energy per unit volume in the material, ΔQ, is proportional to the change in temperature, Δu (in this section only, Δ is the ordinary difference operator with respect to time, not the Laplacian with respect to space). That is,

$$\Delta Q = c_p \rho \Delta u$$

where c_p is the specific heat capacity and ρ is the mass density of the material. Choosing zero energy at absolute zero temperature, this can be rewritten as

$$Q = c_p \rho u.$$

The increase in internal energy in a small spatial region of the material

$$x - \Delta x \leq \xi \leq x + \Delta x$$

over the time period

$$t - \Delta t \leq \tau \leq t + \Delta t$$

is given by

$$c_p \rho \int_{x-\Delta x}^{x+\Delta x} \left[u\left(\xi, t+\Delta t\right) - u\left(\xi, t-\Delta t\right) \right] d\xi = c_p \rho \int_{t-\Delta t}^{t+\Delta t} \int_{x-\Delta x}^{x+\Delta x} \frac{\partial u}{\partial \tau} d\xi d\tau$$

where the fundamental theorem of calculus was used. If no work is done and there are neither heat sources nor sinks, the change in internal energy in the interval $[x-\Delta x, x+\Delta x]$ is accounted for entirely by the flux of heat across the boundaries. By Fourier's law, this is

$$k \int_{t-\Delta t}^{t+\Delta t} \left[\frac{\partial u}{\partial x}(x+\Delta x, \tau) - \frac{\partial u}{\partial x}(x-\Delta x, \tau) \right] d\tau = k \int_{t-\Delta t}^{t+\Delta t} \int_{x-\Delta x}^{x+\Delta x} \frac{\partial^2 u}{\partial \xi^2} d\xi d\tau$$

again by the fundamental theorem of calculus. By conservation of energy,

$$\int_{t-\Delta t}^{t+\Delta t} \int_{x-\Delta x}^{x+\Delta x} \left[c_p \rho u_\tau - k u_{\xi\xi} \right] d\xi d\tau = 0$$

This is true for any rectangle $[t -\Delta t, t + \Delta t] \times [x - \Delta x, x + \Delta x]$. By the fundamental lemma of the calculus of variations, the integrand must vanish identically:

$$c_p \rho u_t - k u_{xx} = 0.$$

Which can be rewritten as:

$$u_t = \frac{k}{c_p \rho} u_{xx},$$

or:

$$\frac{\partial u}{\partial t} = \frac{k}{c_p \rho} \left(\frac{\partial^2 u}{\partial x^2} \right)$$

which is the heat equation, where the coefficient (often denoted α)

$$\alpha = \frac{k}{c_p \rho}$$

is called the thermal diffusivity.

An additional term may be introduced into the equation to account for radiative loss of heat, which depends upon the excess temperature $u = T - T_s$ at a given point compared with the surroundings. At low excess temperatures, the radiative loss is approximately μu, giving a one-dimensional heat-transfer equation of the form

$$\frac{\partial u}{\partial t} = \frac{k}{c_p \rho} \left(\frac{\partial^2 u}{\partial x^2} \right) - \mu u.$$

At high excess temperatures, however, the Stefan–Boltzmann law gives a net radiative heat-loss proportional to $T^4 - T_s^4$, and the above equation is inaccurate. For large excess temperatures, $T^4 - T_s^4 \approx u^4$, giving a high-temperature heat-transfer equation of the form

$$\frac{\partial u}{\partial t} = \alpha \left(\frac{\partial^2 u}{\partial x^2} \right) - mu^4$$

where $m = \epsilon \sigma p / \rho A c_p$. Here, σ is Stefan's constant, ε is a characteristic constant of the material, p is the sectional perimeter of the bar and A is its cross-sectional area. However, using T instead of u gives a better approximation in this case.

Three-dimensional Problem

In the special cases of wave propagation of heat in an isotropic and homogeneous medium in a 3-dimensional space, this equation is

$$\frac{\partial u}{\partial t} = \alpha \nabla^2 u = \alpha \left(\frac{\partial^2 u}{\partial x^2} + \frac{\partial^2 u}{\partial y^2} + \frac{\partial^2 u}{\partial z^2} \right) = \alpha (u_{xx} + u_{yy} + u_{zz})$$

where:

- $u = u(x, y, z, t)$ is temperature as a function of space and time;

- $\dfrac{\partial u}{\partial t}$ is the rate of change of temperature at a point over time;

- u_{xx}, u_{yy}, and u_{zz} are the second spatial derivatives (*thermal conductions*) of temperature in the x, y, and z directions, respectively;

- $\alpha = \dfrac{k}{c_p \rho}$ is the thermal diffusivity, a material-specific quantity depending on the *thermal conductivity k*, the *mass density* ρ, and the *specific heat capacity* c_p.

The heat equation is a consequence of Fourier's law of conduction.

If the medium is not the whole space, in order to solve the heat equation uniquely we also need to specify boundary conditions for u. To determine uniqueness of solutions in the whole space it is necessary to assume an exponential bound on the growth of solutions.

Solutions of the heat equation are characterized by a gradual smoothing of the initial temperature distribution by the flow of heat from warmer to colder areas of an object. Generally, many different states and starting conditions will tend toward the same stable equilibrium. As a consequence, to reverse the solution and conclude something about earlier times or initial conditions from the present heat distribution is very inaccurate except over the shortest of time periods.

The heat equation is the prototypical example of a parabolic partial differential equation.

Using the Laplace operator, the heat equation can be simplified, and generalized to similar equations over spaces of arbitrary number of dimensions, as

$$u_t = \alpha \nabla^2 u = \alpha \Delta u,$$

where the Laplace operator, Δ or ∇^2, the divergence of the gradient, is taken in the spatial variables.

The heat equation governs heat diffusion, as well as other diffusive processes, such as particle diffusion or the propagation of action potential in nerve cells. Although they are not diffusive in nature, some quantum mechanics problems are also governed by a mathematical analog of the heat equation. It also can be used to model some phenomena arising in finance, like the Black–Scholes or the Ornstein-Uhlenbeck processes. The equation, and various non-linear analogues, has also been used in image analysis.

The heat equation is, technically, in violation of special relativity, because its solutions involve instantaneous propagation of a disturbance. The part of the disturbance outside the forward light cone can usually be safely neglected, but if it is necessary to develop a reasonable speed for the transmission of heat, a hyperbolic problem should be considered instead – like a partial differential equation involving a second-order time derivative. Some models of nonlinear heat conduction (which are also parabolic equations) have solutions with finite heat transmission speed.

Internal Heat Generation

The function u above represents temperature of a body. Alternatively, it is sometimes convenient to change units and represent u as the heat density of a medium. Since heat density is proportional to temperature in a homogeneous medium, the heat equation is still obeyed in the new units.

Suppose that a body obeys the heat equation and, in addition, generates its own heat per unit volume (e.g., in watts/litre - W/L) at a rate given by a known function q varying in space and time. Then the heat per unit volume u satisfies an equation

$$\frac{\partial u}{\partial t} = \alpha \left(\frac{\partial^2 u}{\partial x^2} + \frac{\partial^2 u}{\partial y^2} + \frac{\partial^2 u}{\partial z^2} \right) + \frac{1}{c_p \rho} q.$$

For example, a tungsten light bulb filament generates heat, so it would have a positive nonzero value for q when turned on. While the light is turned off, the value of q for the tungsten filament would be zero.

Solving the Heat Equation using Fourier Series

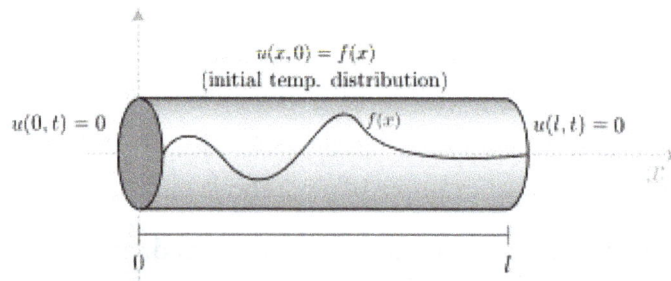

Idealized physical setting for heat conduction in a rod with homogeneous boundary conditions.

The following solution technique for the heat equation was proposed by Joseph Fourier in his treatise *Théorie analytique de la chaleur*, published in 1822. Let us consider the heat equation for one space variable. This could be used to model heat conduction in a rod. The equation is

$$u_t = \alpha u_{xx}$$

where $u = u(x, t)$ is a function of two variables x and t. Here

- x is the space variable, so $x \in [0, L]$, where L is the length of the rod.

- t is the time variable, so $t \geq 0$.

We assume the initial condition

$$u(x,0) = f(x) \quad \forall x \in [0,L]$$

where the function f is given, and the boundary conditions

$$u(0,t) = 0 = u(L,t) \quad \forall t > 0$$

Let us attempt to find a solution of $u_t = \alpha u_{xx}$ that is not identically zero satisfying the boundary conditions but with the following property: u is a product in which the dependence of u on x, t is separated, that is:

$$u(x,t) = X(x)T(t).$$

This solution technique is called separation of variables. Substituting u back into equation $u_t = \alpha u_{xx}$,

$$\frac{T'(t)}{\alpha T(t)} = \frac{X''(x)}{X(x)}.$$

Since the right hand side depends only on x and the left hand side only on t, both sides are equal to some constant value $-\lambda$. Thus:

$$T'(t) = -\lambda\alpha T(t)$$

and

$$X''(x) = -\lambda X(x).$$

We will now show that nontrivial solutions for above equation for values of $\lambda \leq 0$ cannot occur:

1. Suppose that $\lambda < 0$. Then there exist real numbers B, C such that

$$X(x) = Be^{\sqrt{-\lambda}x} + Ce^{-\sqrt{-\lambda}x}.$$

From above equ. we get $X(0) = 0 = X(L)$ and therefore $B = 0 = C$ which implies u is identically 0.

2. Suppose that $\lambda = 0$. Then there exist real numbers B, C such that $X(x) = Bx + C$. From equation $u(0,t) = 0 = u(L,t) \quad \forall t > 0$ we conclude in the same manner as in 1 that u is identically 0.

3. Therefore, it must be the case that $\lambda > 0$. Then there exist real numbers A, B, C such that

$$T(t) = Ae^{-\lambda at}$$

and

$$X(x) = B\sin(\sqrt{\lambda}\,x) + C\cos(\sqrt{\lambda}\,x).$$

From earlier equ. we get $C = 0$ and that for some positive integer n,

$$\sqrt{\lambda} = n\frac{\pi}{L}.$$

This solves the heat equation in the special case that the dependence of u has the special form $u(x,t) = X(x)T(t)$.

We can show that the solution to above equations is given by

$$u(x,t) = \sum_{n=1}^{\infty} D_n \sin\left(\frac{n\pi x}{L}\right) e^{-\frac{n^2\pi^2 at}{L^2}}$$

where

$$D_n = \frac{2}{L}\int_0^L f(x)\sin\left(\frac{n\pi x}{L}\right)dx.$$

Generalizing the Solution Technique

The solution technique used above can be greatly extended to many other types of equations. The idea is that the operator u_{xx} with the zero boundary conditions can be represented in terms of its eigenvectors. This leads naturally to one of the basic ideas of the spectral theory of linear self-adjoint operators.

Consider the linear operator $\Delta u = u_{xx}$. The infinite sequence of functions

$$e_n(x) = \sqrt{\frac{2}{L}}\sin\left(\frac{n\pi x}{L}\right)$$

for $n \geq 1$ are eigenvectors of Δ. Indeed

$$\Delta e_n = -\frac{n^2\pi^2}{L^2}e_n.$$

Moreover, any eigenvector f of Δ with the boundary conditions $f(0) = f(L) = 0$ is of the form e_n for some $n \geq 1$. The functions e_n for $n \geq 1$ form an orthonormal sequence with respect to a certain inner product on the space of real-valued functions on $[0, L]$. This means

$$\langle e_n, e_m \rangle = \int_0^L e_n(x)e_m^*(x)dx = \delta_{mn}$$

Finally, the sequence $\{e_n\}_{n \in \mathbb{N}}$ spans a dense linear subspace of $L^2((0, L))$. This shows that in effect we have diagonalized the operator Δ.

Heat Conduction in non-homogeneous Anisotropic Media

In general, the study of heat conduction is based on several principles. Heat flow is a form of energy flow, and as such it is meaningful to speak of the time rate of flow of heat into a region of space.

- The time rate of heat flow into a region V is given by a time-dependent quantity $q_t(V)$. We assume q has a density Q, so that

$$q_t(V) = \int_V Q(x,t)dx$$

- Heat flow is a time-dependent vector function $H(x)$ characterized as follows: the time rate of heat flowing through an infinitesimal surface element with area dS and with unit normal vector n is

$$H(x) \cdot n(x)dS$$

Thus the rate of heat flow into V is also given by the surface integral

$$q_t(V) = -\int_{\partial V} H(x) \cdot n(x)dS$$

where $n(x)$ is the outward pointing normal vector at x.

- The Fourier law states that heat energy flow has the following linear dependence on the temperature gradient

$$H(x) = -A(x) \cdot \nabla u(x)$$

where $A(x)$ is a 3×3 real matrix that is symmetric and positive definite.

- By the divergence theorem, the previous surface integral for heat flow into V can be transformed into the volume integral

$$q_t(V) = -\int_{\partial V} H(x) \cdot n(x)dS$$
$$= \int_{\partial V} A(x) \cdot \nabla u(x) \cdot n(x)dS$$
$$= \int_V \sum_{i,j} \partial_{x_i}\left(a_{ij}(x)\partial_{x_j} u(x,t)\right)dx$$

- The time rate of temperature change at x is proportional to the heat flowing into an infinitesimal volume element, where the constant of proportionality is dependent on a constant κ

$$\partial_t u(x,t) = \kappa(x)Q(x,t)$$

Putting these equations together gives the general equation of heat flow:

$$\partial_t u(x,t) = \kappa(x)\sum_{i,j}\partial_{x_i}\left(a_{ij}(x)\partial_{x_j}u(x,t)\right)$$

Remarks

- The coefficient $\kappa(x)$ is the inverse of specific heat of the substance at $x \times$ density of the substance at x: $\kappa = 1/(\rho c_p)$.

- In the case of an isotropic medium, the matrix A is a scalar matrix equal to thermal conductivity k.

- In the anisotropic case where the coefficient matrix A is not scalar and/or if it depends on x, then an explicit formula for the solution of the heat equation can seldom be written down. Though, it is usually possible to consider the associated abstract Cauchy problem and show that it is a well-posed problem and/or to show some qualitative properties (like preservation of positive initial data, infinite speed of propagation, convergence toward an equilibrium, smoothing properties). This is usually done by one-parameter semigroups theory: for instance, if A is a symmetric matrix, then the elliptic operator defined by

$$Au(x) := \sum_{i,j}\partial_{x_i}a_{ij}(x)\partial_{x_j}u(x)$$

is self-adjoint and dissipative, thus by the spectral theorem it generates a one-parameter semigroup.

Fundamental Solutions

A fundamental solution, also called a *heat kernel*, is a solution of the heat equation corresponding to the initial condition of an initial point source of heat at a known position. These can be used to find a general solution of the heat equation over certain domains.

In one variable, the Green's function is a solution of the initial value problem

$$\begin{cases} u_t(x,t) - ku_{xx}(x,t) = 0 & (x,t) \in \mathbf{R} \times (0,\infty) \\ u(x,0) = \delta(x) \end{cases}$$

where δ is the Dirac delta function. The solution to this problem is the fundamental solution

$$\Phi(x,t) = \frac{1}{\sqrt{4\pi kt}}\exp\left(-\frac{x^2}{4kt}\right).$$

One can obtain the general solution of the one variable heat equation with initial condition $u(x, 0) = g(x)$ for $-\infty < x < \infty$ and $0 < t < \infty$ by applying a convolution:

$$u(x,t) = \int \Phi(x-y,t)g(y)dy.$$

In several spatial variables, the fundamental solution solves the analogous problem

$$\begin{cases} u_t(\mathbf{x},t) - k\sum_{i=1}^{n} u_{x_i x_i}(\mathbf{x},t) = 0 & (\mathbf{x},t) \in \mathbf{R}^n \times (0,\infty) \\ u(\mathbf{x},0) = \delta(\mathbf{x}) \end{cases}$$

The n-variable fundamental solution is the product of the fundamental solutions in each variable; i.e.,

$$\Phi(\mathbf{x},t) = \Phi(x_1,t)\Phi(x_2,t)\ldots\Phi(x_n,t) = \frac{1}{\sqrt{(4\pi kt)^n}} \exp\left(-\frac{\mathbf{x}\cdot\mathbf{x}}{4kt}\right).$$

The general solution of the heat equation on $\mathbf{R}^n$ is then obtained by a convolution, so that to solve the initial value problem with $u(\mathbf{x}, 0) = g(\mathbf{x})$, one has

$$u(\mathbf{x},t) = \int_{\mathbf{R}^n} \Phi(\mathbf{x}-y,t)g(y)dy.$$

The general problem on a domain Ω in $\mathbf{R}^n$ is

$$\begin{cases} u_t(\mathbf{x},t) - k\sum_{i=1}^{n} u_{x_i x_i}(\mathbf{x},t) = 0 & (\mathbf{x},t) \in \Omega \times (0,\infty) \\ u(\mathbf{x},0) = g(\mathbf{x}) & \mathbf{x} \in \Omega \end{cases}$$

with either Dirichlet or Neumann boundary data. A Green's function always exists, but unless the domain Ω can be readily decomposed into one-variable problems, it may not be possible to write it down explicitly. Other methods for obtaining Green's functions include the method of images, separation of variables, and Laplace transforms.

Some Green's Function Solutions in 1D

A variety of elementary Green's function solutions in one-dimension are recorded here; many others are available elsewhere. In some of these, the spatial domain is $(-\infty,\infty)$. In others, it is the semi-infinite interval $(0,\infty)$ with either Neumann or Dirichlet boundary conditions. One further variation is that some of these solve the inhomogeneous equation

$$u_t = ku_{xx} + f.$$

where f is some given function of x and t.

Homogeneous Heat Equation

Initial value problem on $(-\infty,\infty)$

$$\begin{cases} u_t = ku_{xx} & (x,t) \in \mathbb{R} \times (0,\infty) \\ u(x,0) = g(x) & IC \end{cases}$$

$$u(x,t) = \frac{1}{\sqrt{4\pi kt}} \int_{-\infty}^{\infty} \exp\left(-\frac{(x-y)^2}{4kt}\right) g(y)dy$$

Comment. This solution is the convolution with respect to the variable x of the fundamental solution

$$\Phi(x,t) := \frac{1}{\sqrt{4\pi kt}} \exp\left(-\frac{x^2}{4kt}\right),$$

and the function $g(x)$. (The Green's function number of the fundamental solution is X00.) Therefore, according to the general properties of the convolution with respect to differentiation, $u = g * \Phi$ is a solution of the same heat equation, for

$$\left(\partial_t - k\partial_x^2\right)(\Phi * g) = \left[\left(\partial_t - k\partial_x^2\right)\Phi\right] * g = 0.$$

Moreover,

$$\Phi(x,t) = \frac{1}{\sqrt{t}}\Phi\left(\frac{x}{\sqrt{t}}\right)$$

$$\int_{-\infty}^{\infty} \Phi(x,t)dx = 1,$$

so that, by general facts about approximation to the identity, $\Phi(\cdot, t) * g \to g$ as $t \to 0$ in various senses, according to the specific g. For instance, if g is assumed bounded and continuous on $\mathbb{R}$ then $\Phi(\cdot, t) * g$ converges uniformly to g as $t \to 0$, meaning that $u(x, t)$ is continuous on $\mathbb{R} \times [0, \infty)$ with $u(x, 0) = g(x)$.

Initial value problem on $(0,\infty)$ with homogeneous Dirichlet boundary conditions

$$\begin{cases} u_t = ku_{xx} & (x,t) \in [0,\infty) \times (0,\infty) \\ u(x,0) = g(x) & IC \\ u(0,t) = 0 & BC \end{cases}$$

$$u(x,t) = \frac{1}{\sqrt{kt}} \int_0^{\infty} \left[\exp\left(-\frac{(x-y)^2}{4kt}\right) - \exp\left(-\frac{(x+y)^2}{4kt}\right)\right] g(y)dy$$

Comment. This solution is obtained from the preceding formula as applied to the data $g(x)$ suitably extended to $\mathbb{R}$, so as to be an odd function, that is, letting $g(-x) := -g(x)$ for all x. Correspondingly, the solution of the initial value problem on $(-\infty,\infty)$ is an odd function with respect to the variable x for all values of t, and in particular it satisfies the homogeneous Dirichlet boundary conditions $u(0, t) = 0$. The Green's function number of this solution is X10.

Initial value problem on $(0,\infty)$ with homogeneous Neumann boundary conditions

$$\begin{cases} u_t = ku_{xx} & (x,t) \in [0,\infty) \times (0,\infty) \\ u(x,0) = g(x) & IC \\ u_x(0,t) = 0 & BC \end{cases}$$

$$u(x,t) = \frac{1}{\sqrt{4\pi kt}} \int_0^\infty \left[\exp\left(-\frac{(x-y)^2}{4kt} \right) + \exp\left(-\frac{(x+y)^2}{4kt} \right) \right] g(y) dy$$

Comment. This solution is obtained from the first solution formula as applied to the data *g(x)* suitably extended to R so as to be an even function, that is, letting $g(-x) := g(x)$ for all *x*. Correspondingly, the solution of the initial value problem on R is an even function with respect to the variable *x* for all values of *t*> 0, and in particular, being smooth, it satisfies the homogeneous Neumann boundary conditions $u_x(0, t) = 0$. The Green's function number of this solution is X20.

Problem on (0,∞) with homogeneous initial conditions and non-homogeneous Dirichlet boundary conditions

$$\begin{cases} u_t = ku_{xx} & (x,t) \in [0,\infty) \times (0,\infty) \\ u(x,0) = 0 & IC \\ u(0,t) = h(t) & BC \end{cases}$$

$$u(x,t) = \int_0^t \frac{x}{\sqrt{4\pi k(t-s)^3}} \exp\left(-\frac{x^2}{4k(t-s)} \right) h(s) ds, \qquad \forall x > 0$$

Comment. This solution is the convolution with respect to the variable *t* of

$$\psi(x,t) := -2k\partial_x \Phi(x,t) = \frac{x}{\sqrt{4\pi kt^3}} \exp\left(-\frac{x^2}{4kt} \right)$$

and the function *h(t)*. Since Φ(x, t) is the fundamental solution of

$$\partial_t - k\partial_x^2,$$

the function ψ(x, t) is also a solution of the same heat equation, and so is $u := \psi * h$, thanks to general properties of the convolution with respect to differentiation. Moreover,

$$\psi(x,t) = \frac{1}{x^2} \psi\left(1, \frac{t}{x^2} \right)$$

$$\int_0^\infty \psi(x,t) dt = 1,$$

so that, by general facts about approximation to the identity, ψ(x, ·) *h → h as x → 0 in various senses, according to the specific *h*. For instance, if *h* is assumed continuous on R with support in [0, ∞)

then $\psi(x, \cdot) * h$ converges uniformly on compacta to h as $x \to 0$, meaning that $u(x, t)$ is continuous on $[0, \infty) \times [0, \infty)$ with $u(0, t) = h(t)$.

Inhomogeneous Heat Equation

Problem on $(-\infty, \infty)$ homogeneous initial conditions

$$\begin{cases} u_t = k u_{xx} + f(x,t) & (x,t) \in \mathbb{R} \times (0,\infty) \\ u(x,0) = 0 & IC \end{cases}$$

$$u(x,t) = \int_0^t \int_{-\infty}^\infty \frac{1}{\sqrt{4\pi k(t-s)}} \exp\left(-\frac{(x-y)^2}{4k(t-s)}\right) f(y,s)\,dy\,ds$$

Comment. This solution is the convolution in $\mathbb{R}^2$, that is with respect to both the variables x and t, of the fundamental solution

$$\Phi(x,t) := \frac{1}{\sqrt{4\pi kt}} \exp\left(-\frac{x^2}{4kt}\right)$$

and the function $f(x, t)$, both meant as defined on the whole $\mathbb{R}^2$ and identically 0 for all $t \to 0$. One verifies that

$$\left(\partial_t - k\partial_x^2\right)(\Phi * f) = f,$$

which expressed in the language of distributions becomes

$$\left(\partial_t - k\partial_x^2\right)\Phi = \delta,$$

where the distribution δ is the Dirac's delta function, that is the evaluation at 0.

Problem on $(0, \infty)$ with homogeneous Dirichlet boundary conditions and initial conditions

$$\begin{cases} u_t = k u_{xx} + f(x,t) & (x,t) \in [0,\infty) \times (0,\infty) \\ u(x,0) = 0 & IC \\ u(0,t) = 0 & BC \end{cases}$$

$$u(x,t) = \int_0^t \int_0^\infty \frac{1}{\sqrt{4\pi k(t-s)}} \left(\exp\left(-\frac{(x-y)^2}{4k(t-s)}\right) - \exp\left(-\frac{(x+y)^2}{4k(t-s)}\right)\right) f(y,s)\,dy\,ds$$

Comment. This solution is obtained from the preceding formula as applied to the data $f(x, t)$ suitably extended to $\mathbb{R} \times [0,\infty)$, so as to be an odd function of the variable x, that is, letting $f(-x, t) := -f(x, t)$ for all x and t. Correspondingly, the solution of the inhomogeneous problem on $(-\infty, \infty)$ is an odd function with respect to the variable x for all values of t, and in particular it satisfies the homogeneous Dirichlet boundary conditions $u(0, t) = 0$.

Problem on $(0, \infty)$ with homogeneous Neumann boundary conditions and initial conditions

$$\begin{cases} u_t = ku_{xx} + f(x,t) & (x,t) \in [0,\infty) \times (0,\infty) \\ u(x,0) = 0 & IC \\ u_x(0,t) = 0 & BC \end{cases}$$

$$u(x,t) = \int_0^t \int_0^\infty \frac{1}{\sqrt{4\pi k(t-s)}} \left(\exp\left(-\frac{(x-y)^2}{4k(t-s)} \right) + \exp\left(-\frac{(x+y)^2}{4k(t-s)} \right) \right) f(y,s) \, dy \, ds$$

Comment. This solution is obtained from the first formula as applied to the data $f(x, t)$ suitably extended to $R \times [0,\infty)$, so as to be an even function of the variable x, that is, letting $f(-x, t) := f(x, t)$ for all x and t. Correspondingly, the solution of the inhomogeneous problem on $(-\infty,\infty)$ is an even function with respect to the variable x for all values of t, and in particular, being a smooth function, it satisfies the homogeneous Neumann boundary conditions $u_x(0, t) = 0$.

Examples

Since the heat equation is linear, solutions of other combinations of boundary conditions, inhomogeneous term, and initial conditions can be found by taking an appropriate linear combination of the above Green's function solutions.

For example, to solve

$$\begin{cases} u_t = ku_{xx} + f & (x,t) \in R \times (0,\infty) \\ u(x,0) = g(x) & IC \end{cases}$$

let $u = w + v$ where w and v solve the problems

$$\begin{cases} v_t = kv_{xx} + f, w_t = kw_{xx} & (x,t) \in R \times (0,\infty) \\ v(x,0) = 0, w(x,0) = g(x) & IC \end{cases}$$

Similarly, to solve

$$\begin{cases} u_t = ku_{xx} + f & (x,t) \in [0,\infty) \times (0,\infty) \\ u(x,0) = g(x) & IC \\ u(0,t) = h(t) & BC \end{cases}$$

let $u = w + v + r$ where w, v, and r solve the problems

$$\begin{cases} v_t = kv_{xx} + f, w_t = kw_{xx}, r_t = kr_{xx} & (x,t) \in [0,\infty) \times (0,\infty) \\ v(x,0) = 0, w(x,0) = g(x), r(x,0) = 0 & IC \\ v(0,t) = 0, w(0,t) = 0, r(0,t) = h(t) & BC \end{cases}$$

Mean-value Property for the Heat Equation

Solutions of the heat equations

$$(\partial_t - \Delta)u = 0$$

satisfy a mean-value property analogous to the mean-value properties of harmonic functions, solutions of

$$\Delta u = 0,$$

though a bit more complicated. Precisely, if u solves

$$(\partial_t - \Delta)u = 0$$

and

$$(x,t) + E_\lambda \subset \mathrm{dom}(u)$$

then

$$u(x,t) = \frac{\lambda}{4}\int_{E_\lambda} u(x-y,t-s)\frac{|y|^2}{s^2}\,ds\,dy,$$

where E_λ is a "heat-ball", that is a super-level set of the fundamental solution of the heat equation:

$$E_\lambda := \{(y,s):\Phi(y,s) > \lambda\},$$

$$\Phi(x,t) := (4t\pi)^{-\frac{n}{2}}\exp\left(-\frac{|x|^2}{4t}\right).$$

Notice that

$$\mathrm{diam}(E_\lambda) = o(1)$$

as $\lambda \to \infty$ so the above formula holds for any (x, t) in the (open) set $\mathrm{dom}(u)$ for λ large enough. Conversely, any function u satisfying the above mean-value property on an open domain of $\mathbb{R}^n \times \mathbb{R}$ is a solution of the heat equation. This can be shown by an argument similar to the analogous one for harmonic functions.

Steady-state Heat Equation

The steady-state heat equation is by definition not dependent on time. In other words, it is assumed conditions exist such that:

$$\frac{\partial u}{\partial t} = 0$$

This condition depends on the time constant and the amount of time passed since boundary conditions have been imposed. Thus, the condition is fulfilled in situations in which the *time equilibri-*

um constant is fast enough that the more complex time-dependent heat equation can be approximated by the steady-state case. Equivalently, the steady-state condition exists for all cases in which *enough time has passed* that the thermal field $\boldsymbol{u}$ no longer evolves in time.

In the steady-state case, a spatial thermal gradient may (or may not) exist, but if it does, it does not change in time. This equation therefore describes the end result in all thermal problems in which a source is switched on (for example, an engine started in an automobile), and enough time has passed for all permanent temperature gradients to establish themselves in space, after which these spatial gradients no longer change in time (as again, with an automobile in which the engine has been running for long enough). The other (trivial) solution is for all spatial temperature gradients to disappear as well, in which case the temperature become uniform in space, as well.

The equation is much simpler and can help to understand better the physics of the materials without focusing on the dynamic of the heat transport process. It is widely used for simple engineering problems assuming there is equilibrium of the temperature fields and heat transport, with time.

Steady-state condition:

$$\frac{\partial u}{\partial t} = 0$$

The steady-state heat equation for a volume that contains a heat source (the inhomogeneous case), is the Poisson's equation:

$$-k\nabla^2 u = q$$

where u is the temperature, k is the thermal conductivity and q the heat-flux density of the source.

In electrostatics, this is equivalent to the case where the space under consideration contains an electrical charge.

The steady-state heat equation without a heat source within the volume (the homogeneous case) is the equation in electrostatics for a volume of free space that does not contain a charge. It is described by Laplace's equation:

$$\nabla^2 u = 0$$

Applications

Particle Diffusion

One can model particle diffusion by an equation involving either:

- the volumetric concentration of particles, denoted c, in the case of collective diffusion of a large number of particles, or

- the probability density function associated with the position of a single particle, denoted P.

In either case, one uses the heat equation

$$c_t = D\Delta c,$$

or

$$P_t = D\Delta P.$$

Both c and P are functions of position and time. D is the diffusion coefficient that controls the speed of the diffusive process, and is typically expressed in meters squared over second. If the diffusion coefficient D is not constant, but depends on the concentration c (or P in the second case), then one gets the nonlinear diffusion equation.

Brownian Motion

Let the stochastic process X be the solution of the stochastic differential equation

$$\begin{cases} dX_t = \sqrt{2k}\,dB_t \\ X_0 = 0 \end{cases}$$

where B is the Wiener process (standard Brownian motion). Then the probability density function of X is given at any time t by

$$\frac{1}{\sqrt{4\pi kt}}\exp\left(-\frac{x^2}{4kt}\right)$$

which is the solution of the initial value problem

$$\begin{cases} u_t(x,t) - ku_{xx}(x,t) = 0, & (x,t) \in \mathbb{R}\times(0,+\infty) \\ u(x,0) = \delta(x) \end{cases}$$

where δ is the Dirac delta function.

Schrödinger Equation for a Free Particle

With a simple division, the Schrödinger equation for a single particle of mass m in the absence of any applied force field can be rewritten in the following way:

$$\psi_t = \frac{i\hbar}{2m}\Delta\psi,$$

where i is the imaginary unit, $\hbar$ is the reduced Planck's constant, and ψ is the wave function of the particle.

This equation is formally similar to the particle diffusion equation, which one obtains through the following transformation:

$$c(\mathrm{R},t) \rightarrow \psi(\mathrm{R},t)$$

$$D \rightarrow \frac{i\hbar}{2m}$$

Applying this transformation to the expressions of the Green functions determined in the case of particle diffusion yields the Green functions of the Schrödinger equation, which in turn can be used to obtain the wave function at any time through an integral on the wave function at $t = 0$:

$$\psi(\mathrm{R},t) = \int \psi(\mathrm{R}^0, t=0) G(\mathrm{R}-\mathrm{R}^0, t) dR_x^0 \, dR_y^0 \, dR_z^0,$$

with

$$G(\mathrm{R},t) = \left(\frac{m}{2\pi i\hbar t}\right)^{3/2} e^{-\frac{\mathrm{R}^2 m}{2i\hbar t}}.$$

Remark: this analogy between quantum mechanics and diffusion is a purely formal one. Physically, the evolution of the wave function satisfying Schrödinger's equation might have an origin other than diffusion.

Thermal Diffusivity in Polymers

A direct practical application of the heat equation, in conjunction with Fourier theory, in spherical coordinates, is the prediction of thermal transfer profiles and the measurement of the thermal diffusivity in polymers (Unsworth and Duarte). This dual theoretical-experimental method is applicable to rubber, various other polymeric materials of practical interest, and microfluids. These authors derived an expression for the temperature at the center of a sphere T_C

$$\frac{T_C - T_S}{T_0 - T_S} = 2\sum_{n=1}^{\infty}(-1)^{n+1} \exp\left(-\frac{n^2\pi^2\alpha t}{L^2}\right)$$

where T_0 is the initial temperature of the sphere and T_S the temperature at the surface of the sphere, of radius L. This equation has also found applications in protein energy transfer and thermal modeling in biophysics.

Further Applications

The heat equation arises in the modeling of a number of phenomena and is often used in financial mathematics in the modeling of options. The famous Black–Scholes option pricing model's differential equation can be transformed into the heat equation allowing relatively easy solutions from a familiar body of mathematics. Many of the extensions to the simple option models do not have closed form solutions and thus must be solved numerically to obtain a modeled option price. The equation describing pressure diffusion in a porous medium is identical in form with the heat equation. Diffusion problems dealing with Dirichlet, Neumann and Robin boundary conditions have closed form analytic solutions (Thambynayagam 2011). The heat equation is also widely used in image analysis (Perona & Malik 1990) and in machine-learning as the driving theory behind scale-space or graph Laplacian methods. The heat equation can be efficiently solved numerically using the implicit Crank–Nicolson method of (Crank & Nicolson 1947). This method can be extended to many of the models with no closed form solution.

An abstract form of heat equation on manifolds provides a major approach to the Atiyah–Singer index theorem, and has led to much further work on heat equations in Riemannian geometry.

Hyperbolic Partial Differential Equation

In mathematics, a hyperbolic partial differential equation of order n is a partial differential equation (PDE) that, roughly speaking, has a well-posed initial value problem for the first $n-1$ derivatives. More precisely, the Cauchy problem can be locally solved for arbitrary initial data along any non-characteristic hypersurface. Many of the equations of mechanics are hyperbolic, and so the study of hyperbolic equations is of substantial contemporary interest. The model hyperbolic equation is the wave equation. In one spatial dimension, this is

$$\frac{\partial^2 u}{\partial t^2} = c \frac{\partial^2 u}{\partial x^2}$$

The equation has the property that, if u and its first time derivative are arbitrarily specified initial data on the line $t = 0$ (with sufficient smoothness properties), then there exists a solution for all time t.

The solutions of hyperbolic equations are "wave-like." If a disturbance is made in the initial data of a hyperbolic differential equation, then not every point of space feels the disturbance at once. Relative to a fixed time coordinate, disturbances have a finite propagation speed. They travel along the characteristics of the equation. This feature qualitatively distinguishes hyperbolic equations from elliptic partial differential equations and parabolic partial differential equations. A perturbation of the initial (or boundary) data of an elliptic or parabolic equation is felt at once by essentially all points in the domain.

Although the definition of hyperbolicity is fundamentally a qualitative one, there are precise criteria that depend on the particular kind of differential equation under consideration. There is a well-developed theory for linear differential operators, due to Lars Gårding, in the context of microlocal analysis. Nonlinear differential equations are hyperbolic if their linearizations are hyperbolic in the sense of Gårding. There is a somewhat different theory for first order systems of equations coming from systems of conservation laws.

Definition

A partial differential equation is hyperbolic at a point P provided that the Cauchy problem is uniquely solvable in a neighborhood of P for any initial data given on a non-characteristic hypersurface passing through P. Here the prescribed initial data consist of all (transverse) derivatives of the function on the surface up to one less than the order of the differential equation.

Examples

By a linear change of variables, any equation of the form

$$A\frac{\partial^2 u}{\partial x^2} + 2B\frac{\partial^2 u}{\partial x \partial y} + C\frac{\partial^2 u}{\partial y^2} + (\text{lower order derivative terms}) = 0$$

with

$$B^2 - AC > 0$$

can be transformed to the wave equation, apart from lower order terms which are inessential for the qualitative understanding of the equation. This definition is analogous to the definition of a planar hyperbola.

The one-dimensional wave equation:

$$\frac{\partial^2 u}{\partial t^2} - c^2\frac{\partial^2 u}{\partial x^2} = 0$$

is an example of a hyperbolic equation. The two-dimensional and three-dimensional wave equations also fall into the category of hyperbolic PDE. This type of second-order hyperbolic partial differential equation may be transformed to a hyperbolic system of first-order differential equations.

Hyperbolic System of Partial Differential Equations

The following is a system of s first order partial differential equations for s unknown functions $\vec{u} = (u_1, \ldots, u_s)$, $\vec{u} = \vec{u}(\vec{x}, t)$, where $\vec{x} \in \mathbb{R}^d$:

$$(*) \quad \frac{\partial \vec{u}}{\partial t} + \sum_{j=1}^{d} \frac{\partial}{\partial x_j}\overrightarrow{f^j}(\vec{u}) = 0,$$

where $\overrightarrow{f^j} \in C^1(\mathbb{R}^s, \mathbb{R}^s), j = 1, \ldots, d$ are once continuously differentiable functions, nonlinear in general.

Next, for each $\overrightarrow{f^j}$ a Jacobian matrix $s \times s$ is defined

$$A^j := \begin{pmatrix} \dfrac{\partial f_1^j}{\partial u_1} & \cdots & \dfrac{\partial f_1^j}{\partial u_s} \\ \vdots & \ddots & \vdots \\ \dfrac{\partial f_s^j}{\partial u_1} & \cdots & \dfrac{\partial f_s^j}{\partial u_s} \end{pmatrix}, \text{ for } j = 1, \ldots, d.$$

The system $(*)$ is hyperbolic if for all $\alpha_1, \ldots, \alpha_d \in \mathbb{R}$ the matrix $A := \alpha_1 A^1 + \cdots + \alpha_d A^d$ has only rea-leigenvalues and is diagonalizable.

If the matrix A has *sdistinct* real eigenvalues, it follows that it's diagonalizable. In this case the system (*) is called strictly hyperbolic.

If the matrix A is symmetric, it follows that it's diagonalizable and the eigenvalues are real. In this case the system (*) is called symmetric hyperbolic.

Hyperbolic System and Conservation Laws

There is a connection between a hyperbolic system and a conservation law. Consider a hyperbolic system of one partial differential equation for one unknown function $u = u(\vec{x}, t)$. Then the system (*) has the form

$$(**) \quad \frac{\partial u}{\partial t} + \sum_{j=1}^{d} \frac{\partial}{\partial x_j} f^j(u) = 0.$$

Here, u can be interpreted as a quantity that moves around according to the flux given by $\vec{f} = (f^1, \ldots, f^d)$. To see that the quantity u is conserved, integrate (**) over a domain Ω

$$\int_{\Omega} \frac{\partial u}{\partial t} d\Omega + \int_{\Omega} \nabla \cdot \vec{f}(u) d\Omega = 0.$$

If u and $\vec{f}$ are sufficiently smooth functions, we can use the divergence theorem and change the order of the integration and $\partial / \partial t$ to get a conservation law for the quantity u in the general form

$$\frac{d}{dt} \int_{\Omega} u \, d\Omega + \int_{\partial\Omega} \vec{f}(u) \cdot \vec{n} \, d\Gamma = 0,$$

which means that the time rate of change of u in the domain Ω is equal to the net flux of u through its boundary $\partial\Omega$. Since this is an equality, it can be concluded that u is conserved within Ω.

Wave Equation

The wave equation is an important second-order linear partial differential equation for the description of waves—as they occur in physics—such as sound waves, light waves and water waves. It arises in fields like acoustics, electromagnetics, and fluid dynamics.

Historically, the problem of a vibrating string such as that of a musical instrument was studied by Jean le Rond d'Alembert, Leonhard Euler, Daniel Bernoulli, and Joseph-Louis Lagrange. In 1746, d'Alembert discovered the one-dimensional wave equation, and within ten years Euler discovered the three-dimensional wave equation.

A pulse traveling through a string with fixed endpoints as modeled by the wave equation.

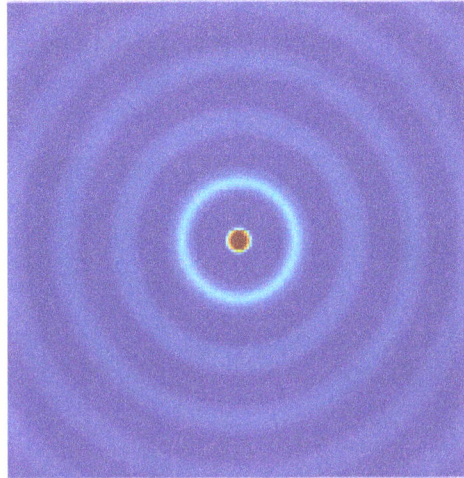

Spherical waves coming from a point source.

Introduction

The wave equation is a hyperbolic partial differential equation. It typically concerns a time variable t, one or more spatial variables $x_1, x_2, ..., x_n$, and a scalar function $u = u (x_1, x_2, ..., x_n; t)$, whose values could model, for example, the mechanical displacement of a wave. The wave equation for u is

$$\frac{\partial^2 u}{\partial t^2} = c^2 \nabla^2 u$$

where ∇^2 is the (spatial) Laplacian and c is a fixed constant.

Solutions of this equation describe propagation of disturbances out from the region at a fixed speed in one or in all spatial directions, as do physical waves from plane or localized sources; the constant c is identified with the propagation speed of the wave. This equation is linear. Therefore, the sum of any two solutions is again a solution: in physics this property is called the superposition principle.

The wave equation alone does not specify a physical solution; a unique solution is usually obtained by setting a problem with further conditions, such as initial conditions, which prescribe the amplitude and phase of the wave. Another important class of problems occurs in enclosed spaces specified by boundary conditions, for which the solutions represent standing waves, or harmonics, analogous to the harmonics of musical instruments.

The wave equation, and modifications of it, are also found in elasticity, quantum mechanics, plasma physics and general relativity.

Scalar Wave Equation in One Space Dimension

The wave equation in one space dimension can be written as follows:

$$\frac{\partial^2 u}{\partial t^2} = c^2 \frac{\partial^2 u}{\partial x^2}.$$

This equation is typically described as having only one space dimension "x", because the only other independent variable is the time "t". Nevertheless, the dependent variable "u" may represent a second space dimension, if, for example, the displacement "u" takes place in y-direction, as in the case of a string that is located in the x-y plane.

French scientist Jean-Baptiste le Rond d'Alembert (b. 1717) discovered the wave equation in one space dimension.

Derivation of the Wave Equation

The wave equation in one space dimension can be derived in a variety of different physical settings. Most famously, it can be derived for the case of a string that is vibrating in a two-dimensional plane, with each of its elements being pulled in opposite directions by the force of tension.

Another physical setting for derivation of the wave equation in one space dimension utilizes Hooke's Law. In the theory of elasticity, Hooke's Law is an approximation for certain materials, stating that the amount by which a material body is deformed (the strain) is linearly related to the force causing the deformation (the stress).

From Hooke's Law

The wave equation in the one-dimensional case can be derived from Hooke's Law in the following way: Imagine an array of little weights of mass m interconnected with massless springs of length h. The springs have a spring constant of k:

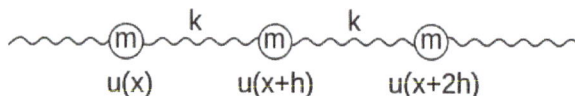

Here the dependent variable $u(x)$ measures the distance from the equilibrium of the mass situated at x, so that $u(x)$ essentially measures the magnitude of a disturbance (i.e. strain) that is traveling in an elastic material. The forces exerted on the mass m at the location $x+h$ are:

$$F_{Newton} = m \cdot a(t) = m \cdot \frac{\partial^2}{\partial t^2} u(x+h,t)$$

$$F_{Hooke} = F_{x+2h} - F_x = k[u(x+2h,t) - u(x+h,t)] - k[u(x+h,t) - u(x,t)]$$

The equation of motion for the weight at the location $x+h$ is given by equating these two forces:

$$\frac{\partial^2}{\partial t^2}u(x+h,t) = \frac{k}{m}[u(x+2h,t) - u(x+h,t) - u(x+h,t) + u(x,t)]$$

where the time-dependence of $u(x)$ has been made explicit.

If the array of weights consists of N weights spaced evenly over the length $L = Nh$ of total mass $M = Nm$, and the total spring constant of the array $K = k/N$ we can write the above equation as:

$$\frac{\partial^2}{\partial t^2}u(x+h,t) = \frac{KL^2}{M}\frac{u(x+2h,t) - 2u(x+h,t) + u(x,t)}{h^2}$$

Taking the limit $N \to \infty$, $h \to 0$ and assuming smoothness one gets:

$$\frac{\partial^2 u(x,t)}{\partial t^2} = \frac{KL^2}{M}\frac{\partial^2 u(x,t)}{\partial x^2}$$

$(KL^2)/M$ is the square of the propagation speed in this particular case.

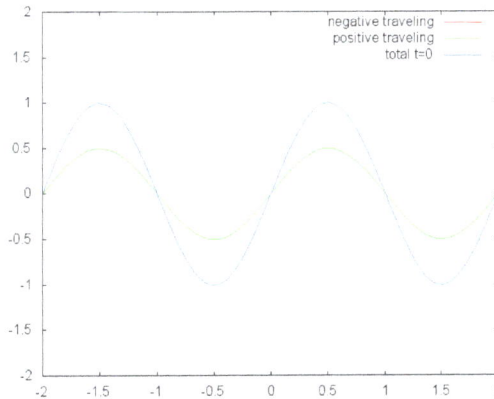

1-d standing wave as a superposition of two waves traveling in opposite directions

Stress Pulse in a Bar

In the case of a stress pulse propagating through a beam the beam acts much like an infinite number of springs in series and can be taken as an extension of the equation derived for Hooke's law. A beam of constant cross section made from a linear elastic material has a stiffness K given by

$$K = \frac{EA}{L}$$

Where A is the cross sectional area and E is the Young's modulus of the material. The wave equation becomes

$$\frac{\partial^2 u(x,t)}{\partial t^2} = \frac{EAL}{M} \frac{\partial^2 u(x,t)}{\partial x^2}$$

AL is equal to the volume of the beam and therefore : $\frac{AL}{M} = \frac{1}{\rho}$ where ρ is the density of the material. The wave equation reduces to

$$\frac{\partial^2 u(x,t)}{\partial t^2} = \frac{E}{\rho} \frac{\partial^2 u(x,t)}{\partial x^2}$$

The speed of a stress wave in a beam is therefore $\sqrt{\frac{E}{\rho}}$

General Solution

Algebraic Approach

The one-dimensional wave equation is unusual for a partial differential equation in that a relatively simple general solution may be found. Defining new variables:

$$\xi = x - ct \quad ; \quad \eta = x + ct$$

changes the wave equation into

$$\frac{\partial^2 u}{\partial \xi \partial \eta} = 0$$

which leads to the general solution

$$u(\xi, \eta) = F(\xi) + G(\eta)$$

or equivalently:

$$u(x,t) = F(x - ct) + G(x + ct)$$

In other words, solutions of the 1D wave equation are sums of a right traveling function F and a left traveling function G. "Traveling" means that the shape of these individual arbitrary functions with respect to x stays constant, however the functions are translated left and right with time at the speed c. This was derived by Jean le Rond d'Alembert.

Another way to arrive at this result is to note that the wave equation may be "factored":

$$\left[\frac{\partial}{\partial t} - c\frac{\partial}{\partial x}\right]\left[\frac{\partial}{\partial t} + c\frac{\partial}{\partial x}\right] u = 0$$

and therefore:

$$\text{either} \quad \frac{\partial u}{\partial t} - c\frac{\partial u}{\partial x} = 0 \quad \text{or} \quad \frac{\partial u}{\partial t} + c\frac{\partial u}{\partial x} = 0$$

These last two equations are advection equations, one left traveling and one right, both with constant speed c.

For an initial value problem, the arbitrary functions F and G can be determined to satisfy initial conditions:

$$u(x,0) = f(x)$$

$$u_t(x,0) = g(x)$$

The result is d'Alembert's formula:

$$u(x,t) = \frac{f(x-ct) + f(x+ct)}{2} + \frac{1}{2c} \int_{x-ct}^{x+ct} g(s)ds$$

In the classical sense if $f(x) \in C^k$ and $g(x) \in C^{k-1}$ then $u(t, x) \in C^k$. However, the waveforms F and G may also be generalized functions, such as the delta-function. In that case, the solution may be interpreted as an impulse that travels to the right or the left.

The basic wave equation is a linear differential equation and so it will adhere to the superposition principle. This means that the net displacement caused by two or more waves is the sum of the displacements which would have been caused by each wave individually. In addition, the behavior of a wave can be analyzed by breaking up the wave into components, e.g. the Fourier transform breaks up a wave into sinusoidal components.

Plane Wave Eigenmodes

Another way to solve for the solutions to the one-dimensional wave equation is to first analyze its frequency eigenmodes. A so-called eigenmode is a solution that oscillates in time with a well-defined *constant* angular frequency ω, with which the temporal part of the wave function for such eigenmode takes a specific form $e^{-i\omega t}$. The rest of the wave function is then only dependent on the spatial variable x, hence amounting to separation of variables. Now writing the wave function as

$$u_\omega(x,t) = e^{-i\omega t} f(x),$$

we can obtain an ordinary differential equation for the spatial part $f(x)$

$$\frac{\partial^2 u_\omega}{\partial t^2} = \frac{\partial^2}{\partial t^2}\left(e^{-i\omega t} f(x)\right) = -\omega^2 e^{-i\omega t} f(x) = c^2 \frac{\partial^2}{\partial x^2}\left(e^{-i\omega t} f(x)\right),$$

Therefore:

$$\frac{d^2}{dx^2} f(x) = -\left(\frac{\omega}{c}\right)^2 f(x),$$

which is precisely an eigenvalue equation for $f(x)$, hence the name eigenmode. It has the well-known plane wave solutions

$$f(x) = A e^{\pm ikx},$$

with wave number $k = \omega / c$.

The total wave function for this eigenmode is then the linear combination

$$u_\omega(x,t) = e^{-i\omega t}\left(Ae^{-ikx} + Be^{ikx}\right) = Ae^{-i(kx+\omega t)} + Be^{i(kx-\omega t)},$$

where complex numbers A, B depend in general on any initial and boundary conditions of the problem.

Eigenmodes are useful in constructing a full solution to the wave equation, because each of them evolves in time trivially with the phase factor $e^{i\omega t}$. so that a full solution can be decomposed into an eigenmode expansion

$$u(x,t) = \int_{-\infty}^{\infty} s(\omega)u_\omega(x,t)d\omega$$

or in terms of the plane waves,

$$
\begin{aligned}
(x,t) &= \int_{-\infty}^{\infty} s_+(\omega)e^{-i(kx+\omega t)}d\omega + \int_{-\infty}^{\infty} s_-(\omega)e^{i(kx-\omega t)}d\omega \\
&= \int_{-\infty}^{\infty} s_+(\omega)e^{-ik(x+ct)}d\omega + \int_{-\infty}^{\infty} s_-(\omega)e^{ik(x-ct)}d\omega \\
&= F(x-ct) + G(x+ct)
\end{aligned}
$$

which is exactly in the same form as in the algebraic approach. Functions $s_\pm(\omega)$ are known as the Fourier component and are determined by initial and boundary conditions. This is a so-called frequency-domain method, alternative to direct time-domain propagations, such as FDTD method, of the wave packet $u(x,t)$, which is complete for representing waves in absence of time dilations. Completeness of the Fourier expansion for representing waves in the presence of time dilations has been challenged by chirp wave solutions allowing for time variation of ω. The chirp wave solutions seem particularly implied by very large but previously inexplicable radar residuals in the flyby anomaly, and differ from the sinusoidal solutions in being receivable at any distance only at proportionally shifted frequencies and time dilations, corresponding to past chirp states of the source.

Scalar Wave Equation in Three Space Dimensions

Swiss mathematician and physicist Leonhard Euler (b. 1707) discovered the wave equation in three space dimensions.

A solution of the initial-value problem for the wave equation in three space dimensions can be obtained from the corresponding solution for a spherical wave. The result can then be also used to obtain the same solution in two space dimensions.

Spherical Waves

The wave equation can be solved using the technique of separation of variables. To obtain a solution with constant frequencies, let us first Fourier transform the wave equation in time as

$$\Psi(\vec{r},t) = \int_{-\infty}^{\infty} \Psi(\vec{r},\omega) e^{-i\omega t} d\omega.$$

So we get,

$$\left(\nabla^2 + \frac{\omega^2}{c^2} \right) \Psi(\vec{r},\omega) = 0$$

This is the Helmholtz equation and can be solved using separation of variables. If spherical coordinates are used to describe a problem, then the solution to the angular part of the Helmholtz equation is given by spherical harmonics and the radial equation now becomes

$$\left[\frac{d^2}{dr^2} + \frac{2}{r} \frac{d}{dr} + k^2 - \frac{l(l+1)}{r^2} \right] f_{lm}(r) = 0$$

Here $k \equiv \dfrac{\omega}{c}$ and the complete solution is now given by

$$\Psi(\vec{r},\omega) = \sum_{lm} \left[A_{lm}^{(1)} h_{lm}^{(1)}(kr) + A_{lm}^{(2)} h_{lm}^{(2)}(kr) \right](r) Y_{lm}(\theta,\phi),$$

where $h_{lm}^{(1)}(kr)$ and $h_{lm}^{(2)}(kr)$ are the spherical Hankel functions. To gain a better understanding of the nature of these spherical waves, let us go back and look at the case when $l=0$. In this case, there is no angular dependence and the amplitude depends only on the radial distance i.e. $\psi(\vec{r},t) \to u(r,t)$.. In this case, the wave equation reduces to

$$\left(\nabla^2 - \frac{1}{c^2} \frac{\partial^2}{\partial t^2} \right) \Psi(\vec{r},t) = 0 \to \left(\frac{\partial^2}{\partial r^2} + \frac{2}{r} \frac{\partial}{\partial r} - \frac{1}{c^2} \frac{\partial^2}{\partial t^2} \right) u(r,t) = 0$$

This equation can be rewritten as

$$\frac{\partial^2 (ru)}{\partial t^2} - c^2 \frac{\partial^2 (ru)}{\partial r^2} = 0;$$

where the quantity ru satisfies the one-dimensional wave equation. Therefore, there are solutions in the form

$$u(r,t) = \frac{1}{r} F(r-ct) + \frac{1}{r} G(r+ct),$$

where F and G are general solutions to the one-dimensional wave equation, and can be interpreted as respectively an outgoing or incoming spherical wave. Such waves are generated by a point source, and they make possible sharp signals whose form is altered only by a decrease in amplitude as r increases. Such waves exist only in cases of space with odd dimensions.

For physical examples of non-spherical wave solutions to the 3D wave equation that do possess angular dependence.

Monochromatic Spherical Wave

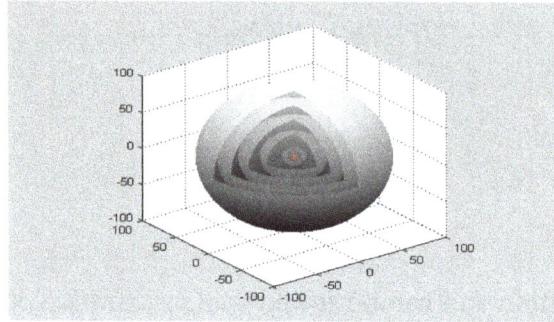

Cut-away of spherical wavefronts, with a wavelength of 10 units, propagating from a point source.

Although the word "monochromatic" is not exactly accurate since it refers to light or electromagnetic radiation with well-defined frequency, the spirit is to discover the eigenmode of the wave equation in three-dimensions. Following the derivation in the previous section on Plane wave eigenmodes, if we again restrict our solutions to spherical waves that oscillate in time with well-defined *constant* angular frequency ω, then the transformed function $ru(r,t)$ has simply plane wave solutions,

$$ru(r,t) = Ae^{i(\omega t \pm kr)},$$

or

$$u(r,t) = \frac{A}{r}e^{i(\omega t \pm kr)}.$$

From this we can observe that the peak intensity of the spherical wave oscillation, characterized as the squared wave amplitude

$$I = |u(r,t)|^2 = \frac{|A|^2}{r^2}.$$

drops at the rate proportional to $1/r^2$, an example of the inverse-square law.

Solution of a General Initial-value Problem

The wave equation is linear in u and it is left unaltered by translations in space and time. Therefore, we can generate a great variety of solutions by translating and summing spherical waves. Let $\phi(\xi,\eta,\zeta)$ be an arbitrary function of three independent variables, and let the spherical wave form F be a delta-function: that is, let F be a weak limit of continuous functions whose integral is unity, but whose support (the region where the function is non-zero) shrinks to the origin. Let a family of spherical waves have center at (ξ, η, ζ), and let r be the radial distance from that point. Thus

$$r^2 = (x - \xi)^2 + (y - \eta)^2 + (z - \zeta)^2.$$

If u is a superposition of such waves with weighting function ϕ, then

$$u(t, x, y, z) = \frac{1}{4\pi c} \iiint \varphi(\xi, \eta, \zeta) \frac{\delta(r - ct)}{r} d\xi \, d\eta \, d\zeta;$$

the denominator $4\pi c$ is a convenience.

From the definition of the delta-function, u may also be written as

$$u(t, x, y, z) = \frac{t}{4\pi} \iint_S \varphi(x + ct\alpha, y + ct\beta, z + ct\gamma) d\omega,$$

where α, β, and γ are coordinates on the unit sphere S, and ω is the area element on S. This result has the interpretation that $u(t, x)$ is t times the mean value of ϕ on a sphere of radius ct centered at x:

$$u(t, x, y, z) = t M_{ct}[\phi].$$

It follows that

$$u(0, x, y, z) = 0, \quad u_t(0, x, y, z) = \phi(x, y, z).$$

The mean value is an even function of t, and hence if

$$v(t, x, y, z) = \frac{\partial}{\partial t} \left(t M_{ct}[\psi] \right),$$

then

$$v(0, x, y, z) = \psi(x, y, z), \quad v_t(0, x, y, z) = 0.$$

These formulas provide the solution for the initial-value problem for the wave equation. They show that the solution at a given point P, given (t, x, y, z) depends only on the data on the sphere of radius ct that is intersected by the light cone drawn backwards from P. It does *not* depend upon data on the interior of this sphere. Thus the interior of the sphere is a lacuna for the solution. This phenomenon is called Huygens' principle. It is true for odd numbers of space dimension, where for one dimension the integration is performed over the boundary of an interval with respect to the Dirac measure. It is not satisfied in even space dimensions. The phenomenon of lacunas has been extensively investigated in Atiyah, Bott and Gårding (1970, 1973).

Scalar Wave Equation in Two Space Dimensions

In two space dimensions, the wave equation is

$$u_{tt} = c^2 \left(u_{xx} + u_{yy} \right).$$

We can use the three-dimensional theory to solve this problem if we regard u as a function in three dimensions that is independent of the third dimension. If

$$u(0, x, y) = 0, \quad u_t(0, x, y) = \phi(x, y),$$

then the three-dimensional solution formula becomes

$$u(t, x, y) = t M_{ct}[\phi] = \frac{t}{4\pi} \iint_S \phi(x + ct\alpha, y + ct\beta) d\omega,$$

where α and β are the first two coordinates on the unit sphere, and $d\omega$ is the area element on the sphere. This integral may be rewritten as a double integral over the disc D with center (x,y) and radius ct:

$$u(t, x, y) = \frac{1}{2\pi c} \iint_D \frac{\phi(x + \xi, y + \eta)}{\sqrt{(ct)^2 - \xi^2 - \eta^2}} d\xi \, d\eta.$$

It is apparent that the solution at (t,x,y) depends not only on the data on the light cone where

$$(x - \xi)^2 + (y - \eta)^2 = c^2 t^2,$$

but also on data that are interior to that cone.

Scalar Wave Equation in General Dimension and Kirchhoff's Formulae

We want to find solutions to $u_{tt} - \Delta u = 0$ for $u : \mathbf{R}^n \times (0, \infty) \to \mathbf{R}$ with $u(x, 0) = g(x)$ and $u_t(x, 0) = h(x)$.

Odd Dimensions

Assume $n \geq 3$ is an odd integer and $g \in C^{m+1}(\mathbf{R}^n)$, $h \in C^m(\mathbf{R}^n)$ for $m = (n+1)/2$. Let $\gamma_n = 1 \cdot 3 \cdot 5 \cdots (n-2)$ and let

$$u(x, t) = \frac{1}{\gamma_n} \left[\partial_t \left(\frac{1}{t} \partial_t \right)^{\frac{n-3}{2}} \left(t^{n-2} \frac{1}{|\partial B_t(x)|} \int_{\partial B_t(x)} g dS \right) + \left(\frac{1}{t} \partial_t \right)^{\frac{n-3}{2}} \left(t^{n-2} \frac{1}{|\partial B_t(x)|} \int_{\partial B_t(x)} h dS \right) \right]$$

then

$$u \in C^2(\mathbf{R}^n \times [0, \infty))$$

$$u_{tt} - \Delta u = 0 \text{ in } \mathbf{R}^n \times (0, \infty)$$

$$\lim_{(x,t) \to (x^0, 0)} u(x, t) = g(x^0)$$

$$\lim_{(x,t) \to (x^0, 0)} u_t(x, t) = h(x^0)$$

Even Dimensions

Assume $n \geq 2$ is an even integer and $g \in C^{m+1}(\mathbf{R}^n)$, $h \in C^m(\mathbf{R}^n)$, for $m = (n+2)/2$. Let $\gamma_n = 2 \cdot 4 \cdot \ldots \cdot n$ and let

$$u(x,t) = \frac{1}{\gamma_n}\left[\partial_t\left(\frac{1}{t}\partial_t\right)^{\frac{n-2}{2}}\left(t^n\frac{1}{|B_t(x)|}\int_{B_t(x)}\frac{g}{(t^2-|y-x|^2)^{\frac{1}{2}}}dy\right)+\left(\frac{1}{t}\partial_t\right)^{\frac{n-2}{2}}\left(t^n\frac{1}{|B_t(x)|}\int_{B_t(x)}\frac{h}{(t^2-|y-x|^2)^{\frac{1}{2}}}dy\right)\right]$$

then

$u \in C^2(\mathbf{R}^n \times [0, \infty))$

$u_{tt} - \Delta u = 0$ in $\mathbf{R}^n \times (0, \infty)$

$$\lim_{(x,t)\to(x^0,0)} u(x,t) = g(x^0)$$

$$\lim_{(x,t)\to(x^0,0)} u_t(x,t) = h(x^0)$$

Problems with Boundaries

The Sturm-Liouville Formulation

A flexible string that is stretched between two points $x = 0$ and $x = L$ satisfies the wave equation for $t > 0$ and $0 < x < L$. On the boundary points, u may satisfy a variety of boundary conditions. A general form that is appropriate for applications is

$$-u_x(t,0) + au(t,0) = 0,$$

$$u_x(t,L) + bu(t,L) = 0,$$

where a and b are non-negative. The case where u is required to vanish at an endpoint is the limit of this condition when the respective a or b approaches infinity. The method of separation of variables consists in looking for solutions of this problem in the special form

$$u(t,x) = T(t)v(x).$$

A consequence is that

$$\frac{T''}{c^2T} = \frac{v''}{v} = -\lambda.$$

The eigenvalue λ must be determined so that there is a non-trivial solution of the boundary-value problem

$$v'' + \lambda v = 0,$$

$$-v'(0) + av(0) = 0, \quad v'(L) + bv(L) = 0.$$

This is a special case of the general problem of Sturm–Liouville theory. If a and b are positive, the eigenvalues are all positive, and the solutions are trigonometric functions. A solution that satisfies square-integrable initial conditions for u and u_t can be obtained from expansion of these functions in the appropriate trigonometric series.

Investigation by Numerical Methods

Approximating the continuous string with a finite number of equidistant mass points one gets the following physical model:

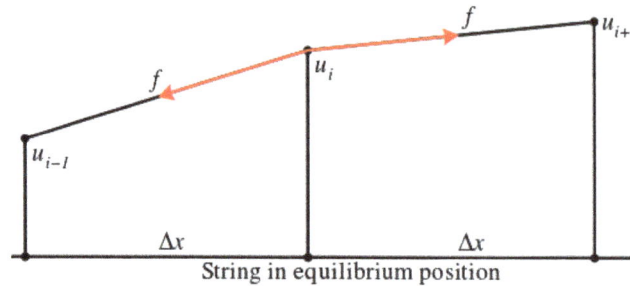

String in equilibrium position

Three consecutive mass points of the discrete model for a string

If each mass point has the mass m, the tension of the string is f, the separation between the mass points is Δx and u_i, $i = 1, ..., n$ are the offset of these n points from their equilibrium points (i.e. their position on a straight line between the two attachment points of the string) the vertical component of the force towards point $i+1$ is

$$\frac{u_{i+1} - u_i}{\Delta x} f$$

and the vertical component of the force towards point $i-1$ is

$$\frac{u_{i-1} - u_i}{\Delta x} f$$

Taking the sum of these two forces and dividing with the mass m one gets for the vertical motion:

$$\ddot{u}_i = \left(\frac{f}{m \Delta x}\right)(u_{i+1} + u_{i-1} - 2u_i)$$

As the mass density is

$$\rho = \frac{m}{\Delta x}$$

this can be written

$$\ddot{u}_i = \left(\frac{f}{\rho \Delta x^2}\right)(u_{i+1} + u_{i-1} - 2u_i)$$

The wave equation is obtained by letting $\Delta x \to 0$ in which case $u_i(t)$ takes the form $u(x, t)$ where $u(x, t)$ is continuous function of two variables, $\ddot{u}_i$ takes the form $\dfrac{\partial^2 u}{\partial t^2}$ and

$$\frac{u_{i+1} + u_{i-1} - 2u_i}{\Delta x^2} \to \frac{\partial^2 u}{\partial x^2}$$

But the discrete formulation of the equation of state with a finite number of mass point is just the suitable one for a numerical propagation of the string motion. The boundary condition

$$u(0, t) = u(L, t) = 0$$

where L is the length of the string takes in the discrete formulation the form that for the outermost points u_1 and u_n the equation of motion are

$$\ddot{u}_1 = \left(\frac{c}{\Delta x}\right)^2 (u_2 - 2u_1)$$

and

$$\ddot{u}_n = \left(\frac{c}{\Delta x}\right)^2 (u_{n-1} - 2u_n)$$

while for $1 < i < n$

$$\ddot{u}_i = \left(\frac{c}{\Delta x}\right)^2 (u_{i+1} + u_{i-1} - 2u_i)$$

where $c = \sqrt{\dfrac{f}{\rho}}$

If the string is approximated with 100 discrete mass points one gets the 100 coupled second order differential equations or equivalently 200 coupled first order differential equations.

Propagating these up to the times

$$\frac{L}{c} k \, 0.05 \quad k = 0, \cdots, 5$$

using an 8th order multistep method the 6 states displayed in figure are found:

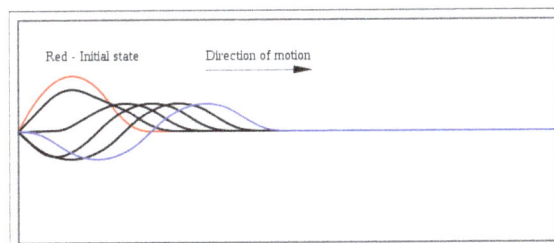

The string at 6 consecutive epochs, the first (red) corresponding to the initial time with the string in rest

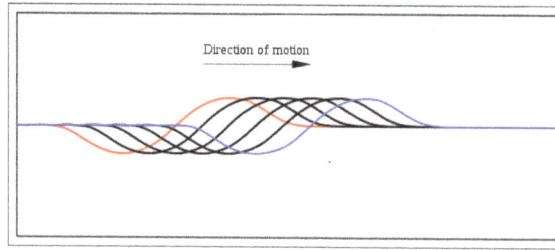

The string at 6 consecutive epochs

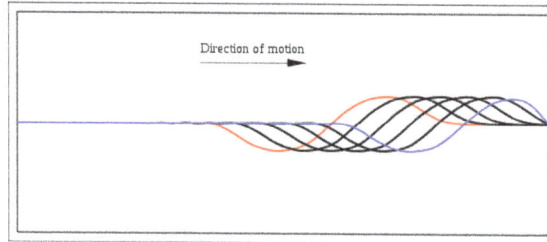

The string at 6 consecutive epochs

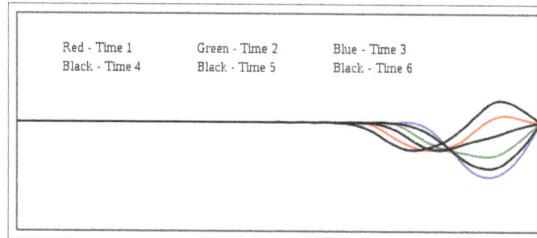

The string at 6 consecutive epochs

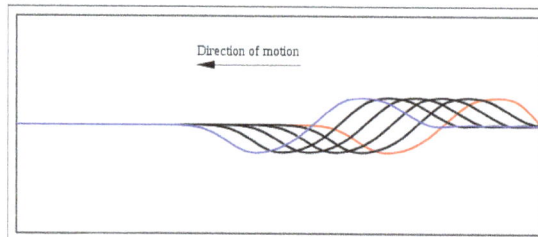

The string at 6 consecutive epochs

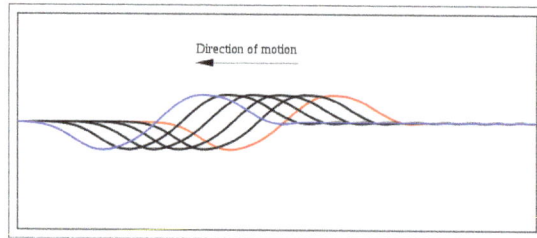

The string at 6 consecutive epochs

The red curve is the initial state at time zero at which the string is "let free" in a predefined shape with all $\dot{u}_i = 0$. The blue curve is the state at time $\frac{L}{c} 0.25$, i.e. after a time that corresponds to the time a wave that is moving with the nominal wave velocity $c = \sqrt{\frac{f}{\rho}}$ would need for one fourth of the length of the string.

Figure displays the shape of the string at the times $\frac{L}{c}\,k\,0.05$ $k = 6,\cdots,11$. The wave travels in direction right with the speed $c = \sqrt{\dfrac{f}{\rho}}$ without being actively constraint by the boundary conditions at the two extremes of the string. The shape of the wave is constant, i.e. the curve is indeed of the form $f(x-ct)$.

Figure displays the shape of the string at the times $\frac{L}{c}\,k\,0.05$ $k = 12,\cdots,17$. The constraint on the right extreme starts to interfere with the motion preventing the wave to raise the end of the string.

Figure displays the shape of the string at the times $\frac{L}{c}\,k\,0.05$ $k = 18,\cdots,23$ when the direction of motion is reversed. The red, green and blue curves are the states at the times $\frac{L}{c}\,k\,0.05$ $k = 18,\cdots,20$ while the 3 black curves correspond to the states at times $\frac{L}{c}\,k\,0.05$ $k = 21,\cdots,23$ with the wave starting to move back towards left.

Figures finally display the shape of the string at the times $\frac{L}{c}\,k\,0.05$ $k = 24,\cdots,29$ and $\frac{L}{c}\,k\,0.05$ $k = 30,\cdots,35$. The wave now travels towards left and the constraints at the end points are not active any more. When finally the other extreme of the string the direction will again be reversed in a way similar to what is displayed in figure.

Several Space Dimensions

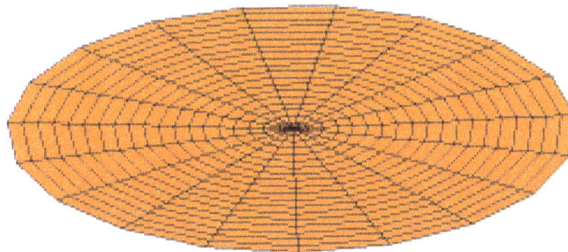

A solution of the wave equation in two dimensions with a zero-displacement
boundary condition along the entire outer edge.

The one-dimensional initial-boundary value theory may be extended to an arbitrary number of space dimensions. Consider a domain D in m-dimensional x space, with boundary B. Then the wave equation is to be satisfied if x is in D and $t > 0$. On the boundary of D, the solution u shall satisfy

$$\frac{\partial u}{\partial n} + au = 0,$$

where n is the unit outward normal to B, and a is a non-negative function defined on B. The case where u vanishes on B is a limiting case for a approaching infinity. The initial conditions are

$$u(0,x) = f(x), \quad u_t(0,x) = g(x),$$

where f and g are defined in D. This problem may be solved by expanding f and g in the eigenfunctions of the Laplacian in D, which satisfy the boundary conditions. Thus the eigenfunction v satisfies

$$\nabla \cdot \nabla v + \lambda v = 0,$$

in D, and

$$\frac{\partial v}{\partial n} + av = 0,$$

on B.

In the case of two space dimensions, the eigenfunctions may be interpreted as the modes of vibration of a drumhead stretched over the boundary B. If B is a circle, then these eigenfunctions have an angular component that is a trigonometric function of the polar angle θ, multiplied by a Bessel function (of integer order) of the radial component. Further details are in Helmholtz equation.

If the boundary is a sphere in three space dimensions, the angular components of the eigenfunctions are spherical harmonics, and the radial components are Bessel functions of half-integer order.

Inhomogeneous Wave Equation in One Dimension

The inhomogeneous wave equation in one dimension is the following:

$$c^2 u_{xx}(x, t) - u_{tt}(x, t) = s(x, t)$$

with initial conditions given by

$$u(x, 0) = f(x)$$

$$u_t(x, 0) = g(x)$$

The function $s(x, t)$ is often called the source function because in practice it describes the effects of the sources of waves on the medium carrying them. Physical examples of source functions include the force driving a wave on a string, or the charge or current density in the Lorenz gauge of electromagnetism.

One method to solve the initial value problem (with the initial values as posed above) is to take advantage of a special property of the wave equation in an odd number of space dimensions, namely that its solutions respect causality. That is, for any point (x_i, t_i), the value of $u(x_i, t_i)$ depends only on the values of $f(x_i+ct_i)$ and $f(x_i-ct_i)$ and the values of the function $g(x)$ between (x_i-ct_i) and (x_i+ct_i). This can be seen in d'Alembert's formula, stated above, where these quantities are the only ones that show up in it. Physically, if the maximum propagation speed is c, then no part of the wave that can't propagate to a given point by a given time can affect the amplitude at the same point and time.

In terms of finding a solution, this causality property means that for any given point on the line being considered, the only area that needs to be considered is the area encompassing all the points that could causally affect the point being considered. Denote the area that casually affects point (x_i, t_i) as R_C. Suppose we integrate the inhomogeneous wave equation over this region.

$$\iint_{R_C} \left(c^2 u_{xx}(x,t) - u_{tt}(x,t) \right) dxdt = \iint_{R_C} s(x,t)dxdt.$$

To simplify this greatly, we can use Green's theorem to simplify the left side to get the following:

$$\int_{L_0+L_1+L_2} \left(-c^2 u_x(x,t)dt - u_t(x,t)dx \right) = \iint_{R_C} s(x,t)dxdt.$$

The left side is now the sum of three line integrals along the bounds of the causality region. These turn out to be fairly easy to compute

$$\int_{x_i-ct_i}^{x_i+ct_i} -u_t(x,0)dx = -\int_{x_i-ct_i}^{x_i+ct_i} g(x)dx.$$

In the above, the term to be integrated with respect to time disappears because the time interval involved is zero, thus $dt = 0$.

For the other two sides of the region, it is worth noting that $x \pm ct$ is a constant, namingly $x_i \pm ct_i$, where the sign is chosen appropriately. Using this, we can get the relation $dx \pm cdt = 0$, again choosing the right sign:

$$\int_{L_1} \left(-c^2 u_x(x,t)dt - u_t(x,t)dx \right) = \int_{L_1} \left(cu_x(x,t)dx + cu_t(x,t)dt \right)$$
$$= c\int_{L_1} du(x,t)$$
$$= cu(x_i, t_i) - cf(x_i + ct_i).$$

And similarly for the final boundary segment:

$$\int_{L_2} \left(-c^2 u_x(x,t)dt - u_t(x,t)dx \right) = -\int_{L_2} \left(cu_x(x,t)dx + cu_t(x,t)dt \right)$$
$$= -c\int_{L_2} du(x,t)$$
$$= cu(x_i, t_i) - cf(x_i - ct_i).$$

Adding the three results together and putting them back in the original integral:

$$\iint_{R_C} s(x,t)dxdt = -\int_{x_i-ct_i}^{x_i+ct_i} g(x)dx + cu(x_i,t_i) - cf(x_i + ct_i) + cu(x_i,t_i) - cf(x_i - ct_i)$$
$$= 2cu(x_i,t_i) - cf(x_i + ct_i) - cf(x_i - ct_i) - \int_{x_i-ct_i}^{x_i+ct_i} g(x)dx$$

Solving for $u(x_i, t_i)$ we arrive at

$$u(x_i, t_i) = \frac{f(x_i + ct_i) + f(x_i - ct_i)}{2} + \frac{1}{2c}\int_{x_i - ct_i}^{x_i + ct_i} g(x)dx + \frac{1}{2c}\int_0^{t_i}\int_{x_i - c(t_i - t)}^{x_i + c(t_i - t)} s(x, t)dxdt.$$

In the last equation of the sequence, the bounds of the integral over the source function have been made explicit. Looking at this solution, which is valid for all choices (x_i, t_i) compatible with the wave equation, it is clear that the first two terms are simply d'Alembert's formula, as stated above as the solution of the homogeneous wave equation in one dimension. The difference is in the third term, the integral over the source.

Other Coordinate Systems

In three dimensions, the wave equation, when written in elliptic cylindrical coordinates, may be solved by separation of variables, leading to the Mathieu differential equation.

Further Generalizations

Elastic Waves

The elastic wave equation in three dimensions describes the propagation of waves in an isotropichomogeneouselastic medium. Most solid materials are elastic, so this equation describes such phenomena as seismic waves in the Earth and ultrasonic waves used to detect flaws in materials. While linear, this equation has a more complex form than the equations given above, as it must account for both longitudinal and transverse motion:

$$\rho\ddot{u} = f + (\lambda + 2\mu)\nabla(\nabla \cdot u) - \mu\nabla \times (\nabla \times u)$$

where:

- λ and μ are the so-called Lamé parameters describing the elastic properties of the medium,

- ρ is the density,

- f is the source function (driving force),

- and u is the displacement vector.

Note that in this equation, both force and displacement are vector quantities. Thus, this equation is sometimes known as the vector wave equation. As an aid to understanding, the reader will observe that if f and $\nabla \cdot u$ are set to zero, this becomes (effectively) Maxwell's equation for the propagation of the electric field E, which has only transverse waves.

Dispersion Relation

In dispersive wave phenomena, the speed of wave propagation varies with the wavelength of the wave, which is reflected by a dispersion relation

$$\omega = \omega(k),$$

where ω is the angular frequency and k is the wavevector describing plane wave solutions. For light waves, the dispersion relation is $\omega = \pm c\,|k|$, but in general, the constant speed c gets replaced by a variable phase velocity:

$$v_p = \frac{\omega(k)}{k}.$$

Second-Order Partial Differential Equations

Classification of PDEs is an important concept because the general theory and methods of solution usually apply only to a given class of equations. Let us first discuss the classification of PDEs involving two independent variables.

Classification with Two Independent Variables

Consider the following general second order linear PDE in two independent variables:

$$A\frac{\partial^2 u}{\partial x^2} + B\frac{\partial^2 u}{\partial x \partial y} + C\frac{\partial^2 u}{\partial y^2} + D\frac{\partial u}{\partial x} + E\frac{\partial u}{\partial y} + Fu + G = 0$$

where A, B, C, D, E, F and G are functions of the independent variables x and y. The above equation may be written in the form

$$Au_{xx} + Bu_{xy} + Cu_{yy} + f(x, y, u_x, u_y, u) = 0$$

where

$$u_x = \frac{\partial u}{\partial x}, u_y = \frac{\partial u}{\partial y}, u_{xx} = \frac{\partial^2 u}{\partial x^2}, u_{xy} = \frac{\partial^2 u}{\partial x \partial y}, u_{yy} = \frac{\partial^2 u}{\partial y^2}$$

Assume that A, B and C are continuous functions of x and y possessing continuous partial derivatives of as high order as necessary.

The classification of PDE is motivated by the classification of second order algebraic equations in two-variables

$$ax^2 + bxy + cy^2 + dx + ey + f = 0$$

We know that the nature of the curves will be decided by the principal part $ax^2 + bxy + cy^2$ i.e., the term containing highest degree. Depending on the sign of the discriminant $b^2 - 4ac$, we classify the curve as follows:

If $b^2 - 4ac > 0$ then the curve traces hyperbola.

If $b^2 - 4ac = 0$ then the curve traces parabola.

If $b^2 - 4ac < 0$ then the curve traces ellipse.

With suitable transformation, we can transform above equation into the following normal form

$$\frac{x^2}{a^2} - \frac{y^2}{b^2} = 1 \,(hyperbola).$$

$$x^2 = y \,(parabola).$$

$$\frac{x^2}{a^2} + \frac{y^2}{b^2} = 1 \,(ellipse).$$

Linear PDE with constant coefficients:Let us first consider the following general linear second order PDE in two independent variables x and y with constant coeffcients:

$$Au_{xx} + Bu_{xy} + Cu_{yy} + Du_x + Eu_y + Fu + G = 0$$

where the coeffcients A, B, C, D, E, F and G are constants. The nature of the above equation is determined by the principal part containing highest partial derivatives i.e.,

$$Lu \equiv Au_{xx} + Bu_{xy} + Cu_{yy}.$$

For classification, we attach a symbol to above equation as P (x, y) = Ax^2 + Bxy + Cy^2 (as if we have replaced x by $\dfrac{\partial}{\partial x}$ and y by $\dfrac{\partial}{\partial y}$. Now depending on the sign of the discriminant (B^2 −4AC),the classification of $Au_{xx} + Bu_{xy} + Cu_{yy} + Du_x + Eu_y + Fu + G = 0$ is done as follows:

B^2 − 4AC > 0 ==⇒ Eq. $Au_{xx} + Bu_{xy} + Cu_{yy} + Du_x + Eu_y + Fu + G = 0$ is hyperbolic

B^2 − 4AC = 0 ==⇒ Eq. $Au_{xx} + Bu_{xy} + Cu_{yy} + Du_x + Eu_y + Fu + G = 0$ is parabolic

B^2 − 4AC < 0 ==⇒ Eq. $Au_{xx} + Bu_{xy} + Cu_{yy} + Du_x + Eu_y + Fu + G = 0$ is elliptic

Linear PDE with variable coefficients: The above classification of Equ. is still valid if the coefficients A, B, C, D, E and F depend on x, y. In this case, the above conditions should be satisfied at each point (x, y) in the region where we want to describe its nature e.g., for elliptic we need to verify

B^2(x, y) − 4A(x, y)C(x, y) < 0

for each (x, y) in the region of interest. Thus, we classify linear PDE with variable coefficients as follows:

B^2(x, y) − 4A(x, y)C(x, y) > 0 at (x, y) ==⇒ Eq. $Au_{xx} + Bu_{xy} + Cu_{yy} + Du_x + Eu_y + Fu + G = 0$ is hyperbolic at (x, y)

B^2(x, y) − 4A(x, y)C(x, y) = 0 at (x, y) ==⇒ Eq. $Au_{xx} + Bu_{xy} + Cu_{yy} + Du_x + Eu_y + Fu + G = 0$ is parabolic at (x, y)

B^2(x, y) − 4A(x, y)C(x, y) < 0 at (x, y) ==⇒ Eq. $Au_{xx} + Bu_{xy} + Cu_{yy} + Du_x + Eu_y + Fu + G = 0$ is elliptic at (x, y)

Note: Eq. $Au_{xx} + Bu_{xy} + Cu_{yy} + Du_x + Eu_y + Fu + G = 0$ is hyperbolic, parabolic, or elliptic depends only on the coefficients of the second derivatives. It has nothing to do with the first-derivative terms, the term in u, or the nonhomogeneous term.

EXAMPLE

1. $u_{xx} + u_{yy} = 0$ (Laplace equation). Here, $A = 1$, $B = 0$, $C = 1$ and $B^2 - 4AC = -4 < 0$. Therefore, it is an elliptic type.

2. $u_t = u_{xx}$ (Heat equation). Here, $A = -1$, $B = 0$, $C = 0$ and $B^2 - 4AC = 0$. Thus, it is of parabolic type.

3. $u_{tt} - u_{xx} = 0$ (Wave equation). In this case, $A = -1$, $B = 0$, $C = 1$ and $B^2 - 4AC = 4 > 0$. Hence, it is of hyperbolic type.

4. $u_{xx} + x u_{yy} = 0$, $x \neq 0$ (Tricomi equation). $B^2 - 4AC = -4x$. Given PDE is hyperbolic for $x < 0$ and elliptic for $x > 0$. This example shows that equations with variable coefficients can change form in the different regions of the domain.

Classification with More than Two Variables

Consider the second-order PDE in general form:

$$\sum_{i=1}^{n}\sum_{j=1}^{n} a_{ij} \frac{\partial^2 u}{\partial x_i \partial x_j} + \sum_{i=1}^{n} b_i \frac{\partial u}{\partial x_i} + cu + d = 0$$

where the coefficients a_{ij}, b_i, c and d are functions of $x = (x_1, x_2,..., x_n)$ alone and $u = u(x_1, x_2,..., x_n)$.

Its principal part is

$$L \equiv \sum_{i=1}^{n}\sum_{j=1}^{n} a_{ij} \frac{\partial^2}{\partial x_i \partial x_j}.$$

It is enough to assume that $A = [a_{ij}]$ is symmetric if not, let $\overline{a}_{ij} = \frac{1}{2}\left(a_{ij} + a_{ji}\right)$ and rewrite

$$L \equiv \sum_{i=1}^{n}\sum_{j=1}^{n} \overline{a}_{ij} \frac{\partial^2}{\partial x_i \partial x_j}.$$

Note that $\dfrac{\partial^2 u}{\partial x_i \partial x_j} = \dfrac{\partial^2 u}{\partial x_j \partial x_i}$. As in two-space dimension, let us attach a quadratic form P with above equation (i.e., replacing $\dfrac{\partial u}{\partial x_i}$ by x_i).

$$P(x_1, x_2,..., x_n) = \sum_{i=1}^{n}\sum_{j=1}^{n} a_{ij} x_i x_j.$$

Since A is a real valued symmetric ($a_{ij} = a_{ji}$) matrix, it is diagonalizable with realeigenvalues $\lambda_1, \lambda_2,..., \lambda_n$ (counted with their multiplicities). In other words, there exists a corresponding set of orthonormal set of n eigenvectors, say $\sigma_1, \sigma_2,..., \sigma_n$ with $R = [\sigma_1, \sigma_2,..., \sigma_n]$ as column vectors such that

$$R^T A R = \begin{bmatrix} \lambda_1 & & & & \\ & \lambda_2 & & O & \\ & & \cdot & & \\ & & & \cdot & \\ & O & & & \cdot \\ & & & & \lambda_n \end{bmatrix} = D$$

We now classify earlier equation depending on sign of eigenvalues of A:

(a) If $\lambda_i > 0 \; \forall i$ or $\lambda_i < 0 \; \forall i$ then earlier equation is elliptic type.

(b) If one or more of the $\lambda_i = 0$ then earlier equation is parabolic type.

(c) If one of the $\lambda_i < 0$ or $\lambda_i > 0$, and all the remaining have opposite sign then earlier equation is said to be of hyperbolic type.

EXAMPLE

1. $\nabla^2 u = u_{xx} + u_{yy} + u_{zz} = 0$. In this case, $\lambda_i = 1 > 0$ for all $i = 1, 2, 3$. It is an elliptic PDE since all eigenvalues are of one sign.

2. It is an easy exercise to check that $u_t - \nabla^2 u = 0$ is of parabolic type.

3. The equation $u_{tt} - \nabla^2 u = 0$ is of hyperbolic type.

EXAMPLE. Classify $u_{x_1 x_1} + 2(1 + cx_2)u_{x_2 x_3} = 0$.

To symmetrize, write it as

$$u_{x1x1} + (1 + cx_2)u_{x2x3} + (1 + cx_2)u_{x3x2} = 0$$

i.e., $\partial_x^T A \partial_x - c\partial_{x_2} = 0$, where

$$A = \begin{bmatrix} 1 & 0 & 0 \\ 0 & 0 & 1 + cx_2 \\ 0 & 1 + cx_2 & 0 \end{bmatrix} \quad \partial_x = \begin{bmatrix} \partial_{x_1} \\ \partial_{x_2} \\ \partial_{x_3} \end{bmatrix}$$

Eigenvalues are $\lambda_1 = 1$, $\lambda_2 = 1 + cx_2$, $\lambda_3 = -(1 + cx_2)$ and normalized eigenvectors

$$\sigma_1 = \begin{bmatrix} 1 \\ 0 \\ 0 \end{bmatrix} \quad \sigma_2 = \begin{bmatrix} 0 \\ 1/\sqrt{2} \\ 1/\sqrt{2} \end{bmatrix} \quad \sigma_3 = \begin{bmatrix} 0 \\ 1/\sqrt{2} \\ -1/\sqrt{2} \end{bmatrix}$$

So

$$R = \begin{bmatrix} 1 & 0 & 0 \\ 0 & 1/\sqrt{2} & 1/\sqrt{2} \\ 0 & 1/\sqrt{2} & -1/\sqrt{2} \end{bmatrix}$$

Note that $R = R^T = R^{-1}$.

$$R^T A R = \begin{bmatrix} 1 & 0 & 0 \\ 0 & 1 + cx_2 & 0 \\ 0 & 0 & -(1 + cx_2) \end{bmatrix} = D$$

Equation is parabolic if $x_2 = -\dfrac{1}{c}\,(c \neq 0)$, hyperbolic if $x_2 > -\dfrac{1}{c}$ and $x_2 < -\dfrac{1}{c}$. For $c = 0$, $\lambda_1 = \lambda_2 = 1$ and $\lambda_3 = -1$, it is hyperbolic type.

Canonical Forms or Normal Forms

By a suitable change of the independent variables we shall show that any equation of the form

$$Au_{xx} + Bu_{xy} + Cu_{yy} + Du_x + Eu_y + Fu + G = 0$$

where A, B, C, D, E, F and G are functions of the variables x and y, can be reduced to a canonical form or normal form. The transformed equation assumes a simple form so that the subsequent analysis of solving the equation will be become easy.

Consider the transformation of the indpendent variables from (x, y) to (ξ, η) given by

$$\xi = \xi(x, y),\, \eta = \eta(x, y)$$

Here, the functions ξ and η are continuously differentiable and the Jacobian

$$J = \frac{\partial(\xi, n)}{\partial(x, y)} = \begin{vmatrix} \xi_x & \xi_y \\ \eta_x & \eta_y \end{vmatrix} = \left(\xi_x \eta_y - \xi_y \eta_x \right) \neq 0$$

in the domain where $Au_{xx} + Bu_{xy} + Cu_{yy} + Du_x + Eu_y + Fu + G = 0$ holds.

Using chain rule, we notice that

$$
\begin{aligned}
u_x &= u_\xi \xi_x + u_\eta \eta_x \\
u_y &= u_\xi \xi_y + u_\eta \eta_y \\
u_{xx} &= u_{\xi\xi}\xi_x^2 + 2u_{\xi\eta}\xi_x\eta_x + u_{\eta\eta}\eta_x^2 + u_\xi \xi_{xx} + u_\eta \eta_{xx} \\
u_{xy} &= u_{\xi\xi}\xi_x\xi_y + u_{\xi\eta}(\xi_x\eta_y + \xi_y\eta_x) + u_{\eta\eta}\eta_x\eta_y + u_\xi \xi_{xy} + u_\eta \eta_{xy} \\
u_{yy} &= u_{\xi\xi}\xi_y^2 + 2u_{\xi\eta}\xi_y\eta_y + u_{\eta\eta}\eta_y^2 + u_\xi \xi_{yy} + u_\eta \eta_{yy}
\end{aligned}
$$

Substituting these expressions into $Au_{xx} + Bu_{xy} + Cu_{yy} + Du_x + Eu_y + Fu + G = 0$, we obtain

$$\bar{A}(\xi_x, \xi_y)u_{\xi\xi} + \bar{B}(\xi_x, \xi_y; \eta_x, \eta_y)u_{\xi\eta} + \bar{C}(\eta_x, \eta_y)u_{\eta\eta} = F(\xi, \eta, u(\xi, \eta), u_\xi(\xi, \eta), u_\eta(\xi, \eta))$$

where

$$
\begin{aligned}
\bar{A}(\xi_x, \xi_y) &= A\xi_x^2 + B\xi_x\xi_y + C\xi_y^2 \\
\bar{B}(\xi_x, \xi_y; \eta_x, \eta_y) &= 2A\xi_x\eta_x + B(\xi_x\eta_y + \xi_y\eta_x) + 2C\xi_y\eta_y \\
\bar{C}(\eta_x, \eta_y) &= A\eta_x^2 + B\eta_x\eta_y + C\eta_y^2.
\end{aligned}
$$

An easy calculation shows that

$$\overline{B}^2 - 4\overline{A}\overline{C} = (\xi_x \eta_y - \xi_y \eta_x)^2 (B^2 - 4AC).$$

The above equation shows that the transformation of the independent variables does not modify the type of PDE.

We shall determine ξ and η so that $\overline{A}(\xi_x, \xi_y) u_{\xi\xi} + \overline{B}(\xi_x, \xi_y; \eta_x, \eta_y) u_{\xi\eta} + \overline{C}(\eta_x, \eta_y) u_{\eta\eta} = F(\xi, \eta, u(\xi, \eta), u_\xi(\xi, \eta), u_\eta(\xi, \eta))$ takes the simplest possible form. We now consider the following cases:

      Case I: $B^2 - 4AC > 0$ (Hyperbolic type)

      Case II: $B^2 - 4AC = 0$ (Parabolic type)

      Case III: $B^2 - 4AC < 0$(Elliptic type)

Case I: Note that $B^2 - 4AC > 0$ implies the equation $A\alpha^2 + B\alpha + C = 0$ has two real and distinct roots, say λ_1 and λ_2. Now, choose ξ and η such that

$$\frac{\partial \xi}{\partial x} = \lambda_1 \frac{\partial \xi}{\partial y} \quad and \quad \frac{\partial \eta}{\partial x} = \lambda_2 \frac{\partial \eta}{\partial y}$$

Then the coefficients of $u_{\xi\xi}$ and $u_{\eta\eta}$ will be zero because

$$\overline{A} = A\xi_x^2 + B\xi_x\xi_y + C\xi_y^2 = (A\lambda_1^2 + B\lambda_1 + C)\xi_y^2 = 0,$$
$$\overline{C} = A\eta_x^2 + B\eta_x\eta_y + C\eta_y^2 = (A\lambda_2^2 + B\lambda_2 + C)\eta_y^2 = 0.$$

Thus, $\overline{B}^2 - 4\overline{A}\overline{C} = (\xi_x \eta_y - \xi_y \eta_x)^2 (B^2 - 4AC)$ reduces to

$$\overline{B}^2 = (B^2 - AC)(\xi_x \eta_y - \xi_y \eta_x)^2 > 0$$

as $B^2 - 4AC > 0$. Note that earlier equation is a first-order linear PDE in ξ and η whose characteristics curves satisfy the first-order ODEs

$$\frac{dy}{dx} + \lambda_i(x, y) = 0, \quad i = 1, 2.$$

Let the family of curves determined by the solution of above equation for $i = 1$ and $i = 2$ be

$$f_1(x, y) = c_1 \text{ and } f_2(x, y) = c_2,$$

respectively. These family of curves are called characteristics curves of PDE earlier equation. With this choice, divide earlier equation throughout by $\overline{B}$ (as $\overline{B} > 0$) and use above equation to obtain

$$\frac{\partial^2 u}{\partial \xi \partial \eta} = \phi(\xi, \eta, u, u_\xi, u_\eta)$$

which is the canonical form of hyperbolic equation.

EXAMPLE. Reduce the equation $u_{xx} = x^2 u_{yy}$ to its canonical form.

Solution: Comparing with earlier equation we find that A = 1, B = 0, C = −x².

The roots of the equations $A\alpha^2 + B\alpha + C = 0$ i.e., $\alpha^2 + x^2 = 0$ are given by $\lambda_1 = \pm x$. The differential equations for the family of characteristics curves are

$$\frac{dy}{dx} \pm x = 0.$$

whose solutions are $y + \frac{1}{2}x^2 = c_1$ and $y - \frac{1}{2}x^2 = c_2$. Choose

$$\xi = y + \frac{1}{2}x^2, \quad \eta = y - \frac{1}{2}x^2.$$

An easy computation shows that

$$u_x = u_\xi \xi_x + u_\eta \eta_x,$$
$$u_{xx} = u_{\xi\xi}\xi_x^2 + 2u_{\xi\eta}\xi_x\eta_x + u_{\eta\eta}\eta_x^2 + u_\xi\xi_{xx} + u_\eta\eta_{xx}$$
$$= u_{\xi\xi}x^2 - 2u_{\xi\eta}x^2 + u_{\eta\eta}x^2 + u_\xi - u_\eta,$$
$$u_{yy} = u_{\xi\xi}\xi_y^2 + 2u_{\xi\eta}\xi_y\eta_y + u_{\eta\eta}\eta_y^2 + u_\xi\xi_{yy} + u_\eta\eta_{yy},$$
$$= u_{\xi\xi} + 2u_{\xi\eta} + u_{\eta\eta}.$$

Substituting these expression in the equation $u_{xx} = x^2 u_{yy}$ yields

$$4x^2 u_{\xi\eta} = (u_\xi - u_\eta)$$
$$\text{or} \quad 4(\xi - \eta)u_{\xi\eta} = \frac{1}{4(\xi - \eta)}(u_\xi - u_\eta)$$
$$\text{or} \quad u_{\xi\eta} = \frac{1}{4(\xi - \eta)}(u_\xi - u_\eta)$$

which is the required canonical form.

CASE II: $B^2 - 4AC = 0 \Rightarrow$ the equation $A\alpha^2 + B\alpha + C = 0$ has two equal roots, say $\lambda_1 = \lambda_2 = \lambda$. Let $f_1(x, y) = c_1$ be the solution of $\frac{dy}{dx} + \lambda(x, y) = 0$. Take $\xi = f_1(x, y)$ and η to be the any function of x and y which is independent of ξ.

As before, $\bar{A}(\xi_x, \xi_y) = 0$ and hence from earlier equation, we obtain $\bar{B} = 0$. Note that $\bar{C}(\eta_x, \eta_y) \neq 0$, otherwise η would be a function of ξ. Dividing earlier equation by $\bar{C}$, the canonical form of above equation is

$$u_{\eta\eta} = \phi(\xi, \eta, u, u_\xi, u_\eta).$$

which is the canonical form of parabolic equation.

EXAMPLE. Reduce the equation $u_{xx} + 2u_{xy} + u_{yy} = 0$ to canonical form.

Solution: In this case, A = 1, B = 2, C = 1. The equation $\alpha^2 + 2\alpha + 1 = 0$ has equal roots $\lambda = -1$. The solution of $\frac{dy}{dx} - 1 = 0$ is $x - y = c_1$. Take $\xi = x - y$. Choose $\eta = x + y$ proceed as in Example 1 to obtain $u_{\eta\eta} = 0$ which is the canonical form of the given PDE.

CASE III: When $B^2 - 4AC < 0$, the roots of $A\alpha^2 + B\alpha + C = 0$ are complex. Following the procedure as in CASE I, we find that

$$u_{\xi\eta} = \phi_1(\xi, \eta, u, u_\xi, u_\eta).$$

The variables ξ, η are infact complex conjugates. To get a real canonical form use the transformation

$$\alpha = \frac{1}{2}(\xi + \eta), \quad \beta = \frac{1}{2i}(\xi - \eta)$$

to obtain

$$u_{\xi\eta} = \frac{1}{4}(u_{\alpha\alpha} + u_{\beta\beta})$$

which follows from the following calculation:

$$
\begin{aligned}
u_\xi &= u_\alpha \alpha_\xi + u_\beta \beta_\xi = \frac{1}{2}u_\alpha + \frac{1}{2i}u_\beta \\
u_{\xi\eta} &= \frac{1}{2}(u_{\alpha\alpha}\alpha_\eta + u_{\alpha\beta}\beta_\eta) + \frac{1}{2i}(u_{\beta\alpha}\alpha_\eta + u_{\beta\beta}\beta_\eta) \\
&= \frac{1}{4}(u_{\alpha\alpha} + u_{\beta\beta}).
\end{aligned}
$$

The desired canonical form is

$$u_{\alpha\alpha} + u_{\beta\beta} = \psi(\alpha, \beta, u(\alpha,\beta), u_\alpha(\alpha,\beta), u_\beta(\alpha,\beta))$$

EXAMPLE. Reduce the equation $u_{xx} + x^2 u_{yy} = 0$ to canonical form.

Solution: In this case, A = 1, B = 0, C = x^2. The roots are $\lambda_1 = ix$, $\lambda_2 = -ix$. Take $\xi = iy + \frac{1}{2}x^2$, $\eta = -iy + \frac{1}{2}x^2$. Then $\alpha = \frac{1}{2}x^2$, $\beta = y$. The canonical form is

$$u_{\alpha\alpha} + u_{\beta\beta} = -\frac{1}{2\alpha}u_\alpha.$$

Superposition Principle

In physics and systems theory, the superposition principle, also known as superposition property, states that, for all linear systems, the net response at a given place and time caused by two or more stimuli is the sum of the responses that would have been caused by each stimulus individually. So that if input *A* produces response *X* and input *B* produces response *Y* then input $(A + B)$ produces response $(X + Y)$.

Superposition of almost plane waves (diagonal lines) from a distant source and waves from the wake of the ducks. Linearity holds only approximately in water and only for waves with small amplitudes relative to their wavelengths.

The homogeneity and additivity properties together are called the superposition principle. A linear function is one that satisfies the properties of superposition. It is defined as

$$F(x_1 + x_2) = F(x_1) + F(x_2) \text{ Additivity}$$

$$F(ax) = aF(x) \text{ Homogeneity}$$

for scalar *a*.

This principle has many applications in physics and engineering because many physical systems can be modeled as linear systems. For example, a beam can be modeled as a linear system where the input stimulus is the load on the beam and the output response is the deflection of the beam. The importance of linear systems is that they are easier to analyze mathematically; there is a large body of mathematical techniques, frequency domainlinear transform methods such as Fourier, Laplace transforms, and linear operator theory, that are applicable. Because physical systems are generally only approximately linear, the superposition principle is only an approximation of the true physical behaviour.

The superposition principle applies to *any* linear system, including algebraic equations, linear differential equations, and systems of equations of those forms. The stimuli and responses could be numbers, functions, vectors, vector fields, time-varying signals, or any other object that satisfies certain axioms. Note that when vectors or vector fields are involved, a superposition is interpreted as a vector sum.

Relation to Fourier Analysis and Similar Methods

By writing a very general stimulus (in a linear system) as the superposition of stimuli of a specific, simple form, often the response becomes easier to compute.

For example, in Fourier analysis, the stimulus is written as the superposition of infinitely many sinusoids. Due to the superposition principle, each of these sinusoids can be analyzed separately, and its individual response can be computed. (The response is itself a sinusoid, with the same frequency as the stimulus, but generally a different amplitude and phase.) According to the superposition principle, the response to the original stimulus is the sum (or integral) of all the individual sinusoidal responses.

As another common example, in Green's function analysis, the stimulus is written as the superposition of infinitely many impulse functions, and the response is then a superposition of impulse responses.

Fourier analysis is particularly common for waves. For example, in electromagnetic theory, ordinary light is described as a superposition of plane waves (waves of fixed frequency, polarization, and direction). As long as the superposition principle holds, the behavior of any light wave can be understood as a superposition of the behavior of these simpler plane waves.

Wave Superposition

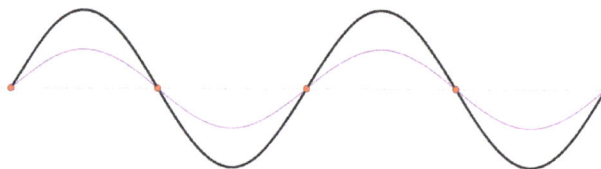

Two waves traveling in opposite directions across the same medium combine linearly. In this, both waves have the same wavelength and the sum of amplitudes results in a standing wave.

Waves are usually described by variations in some parameter through space and time—for example, height in a water wave, pressure in a sound wave, or the electromagnetic field in a light wave. The value of this parameter is called the amplitude of the wave, and the wave itself is a function specifying the amplitude at each point.

In any system with waves, the waveform at a given time is a function of the sources (i.e., external forces, if any, that create or affect the wave) and initial conditions of the system. In many cases (for example, in the classic wave equation), the equation describing the wave is linear. When this is true, the superposition principle can be applied. That means that the net amplitude caused by two or more waves traversing the same space is the sum of the amplitudes that would have been produced by the individual waves separately. For example, two waves traveling towards each other will pass right through each other without any distortion on the other side.

Wave Diffraction Vs. Wave Interference

With regard to wave superposition, Richard Feynman wrote:

No-one has ever been able to define the difference between interference and diffraction satisfactorily. It is just a question of usage, and there is no specific, important physical difference between them. The best we can do is, roughly speaking, is to say that when there are only a few sources, say two, interfering, then the result is usually called interference, but if there is a large number of them, it seems that the word diffraction is more often used.

Other authors elaborate:

The difference is one of convenience and convention. If the waves to be superposed originate from a few coherent sources, say, two, the effect is called interference. On the other hand, if the waves to be superposed originate by subdividing a wavefront into infinitesimal coherent wavelets (sources), the effect is called diffraction. That is the difference between the two phenomena is [a matter] of degree only, and basically they are two limiting cases of superposition effects.

Yet another source concurs:

Inasmuch as the interference fringes observed by Young were the diffraction pattern of the double slit, this chapter [Fraunhofer diffraction] is therefore a continuation of Chapter 8 [Interference]. On the other hand, few opticians would regard the Michelson interferometer as an example of diffraction. Some of the important categories of diffraction relate to the interference that accompanies division of the wavefront, so Feynman's observation to some extent reflects the difficulty that we may have in distinguishing division of amplitude and division of wavefront.

Wave Interference

The phenomenon of interference between waves is based on this idea. When two or more waves traverse the same space, the net amplitude at each point is the sum of the amplitudes of the individual waves. In some cases, such as in noise-cancelling headphones, the summed variation has a smaller amplitude than the component variations; this is called *destructive interference*. In other cases, such as in Line Array, the summed variation will have a bigger amplitude than any of the components individually; this is called *constructive interference*.

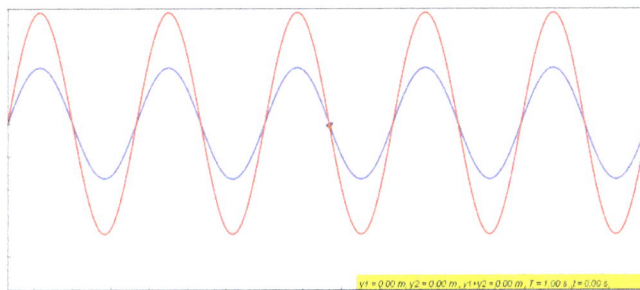

green wave traverse to the right while blue wave traverse left, the net red wave amplitude at each point is the sum of the amplitudes of the individual waves.

combined waveform		
wave 1		
wave 2		
	Two waves in phase	Two waves 180° out of phase

Departures from Linearity

In most realistic physical situations, the equation governing the wave is only approximately linear. In these situations, the superposition principle only approximately holds. As a rule, the accuracy of the approximation tends to improve as the amplitude of the wave gets smaller. For examples of phenomena that arise when the superposition principle does not exactly hold.

Quantum Superposition

In quantum mechanics, a principal task is to compute how a certain type of wave propagates and behaves. The wave is described by a wave function, and the equation governing its behavior is called the Schrödinger equation. A primary approach to computing the behavior of a wave function is to write it as a superposition (called "quantum superposition") of (possibly infinitely many) other wave functions of a certain type—stationary states whose behavior is particularly simple. Since the Schrödinger equation is linear, the behavior of the original wave function can be computed through the superposition principle this way.

The projective nature of quantum-mechanical-state space makes an important difference: it does not permit superposition of the kind that is the topic of the present article. A quantum mechanical state is a *ray* in projective Hilbert space, not a *vector*. The sum of two rays is undefined. To obtain the relative phase, we must decompose or split the ray into components

$$|\psi_i\rangle = \sum_j C_j |\phi_j\rangle,$$

where the $C_j \in C$ and the $|\phi_j\rangle$ belongs to an orthonormal basis set. The equivalence class of $|\psi_i\rangle$ allows a well-defined meaning to be given to the relative phases of the C_j.

There are some likenesses between the superposition presented in the main on this page, and quantum superposition. Nevertheless, on the topic of quantum superposition, Kramers writes: "The principle of [quantum] superposition ... has no analogy in classical physics." According to Dirac: "*the superposition that occurs in quantum mechanics is of an essentially different nature from any occurring in the classical theory* [italics in original]."

Boundary Value Problems

A common type of boundary value problem is (to put it abstractly) finding a function y that satisfies some equation

$$F(y) = 0$$

with some boundary specification

$$G(y) = z$$

For example, in Laplace's equation with Dirichlet boundary conditions, F would be the Laplacian operator in a region R, G would be an operator that restricts y to the boundary of R, and z would be the function that y is required to equal on the boundary of R.

In the case that F and G are both linear operators, then the superposition principle says that a superposition of solutions to the first equation is another solution to the first equation:

$$F(y_1) = F(y_2) = \cdots = 0 \Rightarrow F(y_1 + y_2 + \cdots) = 0$$

while the boundary values superpose:

$$G(y_1) + G(y_2) = G(y_1 + y_2)$$

Using these facts, if a list can be compiled of solutions to the first equation, then these solutions can be carefully put into a superposition such that it will satisfy the second equation. This is one common method of approaching boundary value problems.

Additive State Decomposition

Consider a simple linear system :

$$\dot{x} = Ax + B(u_1 + u_2), x(0) = x_0.$$

By superposition principle, the system can be decomposed into

$$\dot{x}_1 = Ax_1 + Bu_1, x_1(0) = x_0.$$

$$\dot{x}_2 = Ax_2 + Bu_2, x_2(0) = 0.$$

with

$x = x_1 + x_2.$ Superposition principle is only available for linear systems. However, the Additive state decomposition can be applied not only to linear systems but also nonlinear systems. Next, consider a nonlinear system

$$\dot{x} = Ax + B(u_1 + u_2) + \phi(c^T x), x(0) = x_0.$$

where ϕ is a nonlinear function. By the additive state decomposition, the system can be 'additively' decomposed into

$$\dot{x}_1 = Ax_1 + Bu_1 + \phi(y_d), x_1(0) = x_0.$$

$$\dot{x}_2 = Ax_2 + Bu_2 + \phi(c^T x_1 + c^T x_2) - \phi(y_d), x_2(0) = 0.$$

with

$$x = x_1 + x_2.$$

This decomposition can help to simplify controller design.

Other Example Applications

- In electrical engineering, in a linear circuit, the input (an applied time-varying voltage signal) is related to the output (a current or voltage anywhere in the circuit) by a linear transformation. Thus, a superposition (i.e., sum) of input signals will yield the superposition of the responses. The use of Fourier analysis on this basis is particularly common. For another, related technique in circuit analysis.

- In physics, Maxwell's equations imply that the (possibly time-varying) distributions of charges and currents are related to the electric and magnetic fields by a linear transformation. Thus, the superposition principle can be used to simplify the computation of fields which arise from given charge and current distribution. The principle also applies to other linear differential equations arising in physics, such as the heat equation.

- In mechanical engineering, superposition is used to solve for beam and structure deflections of combined loads when the effects are linear (i.e., each load does not affect the results of the other loads, and the effect of each load does not significantly alter the geometry of the structural system). Mode superposition method uses the natural frequencies and mode shapes to characterize the dynamic response of a linear structure.

- In hydrogeology, the superposition principle is applied to the drawdown of two or more water wells pumping in an ideal aquifer.

- In process control, the superposition principle is used in model predictive control.

- The superposition principle can be applied when small deviations from a known solution to a nonlinear system are analyzed by linearization.

- In music, theorist Joseph Schillinger used a form of the superposition principle as one basis of his *Theory of Rhythm* in his *Schillinger System of Musical Composition*.

History

According to Léon Brillouin, the principle of superposition was first stated by Daniel Bernoulli in 1753: "The general motion of a vibrating system is given by a superposition of its proper vibrations." The principle was rejected by Leonhard Euler and then by Joseph Lagrange. Later it became accepted, largely through the work of Joseph Fourier.

Superposition Principle and Wellposedness

A very important fact concerning linear PDEs is the superposition principle, which is stated below.

A linear PDE can be written in the form

$$L[u] = f,$$

where L[u] denotes a linear combination of u and some of its partial derivatives, with coefficients which are given functions of the independent variables.

DEFINITION. (Superposition principle) Let u_1 be a solution of the linear PDE

$$L[u] = f_1$$

and let u_2 be a solution of the linear PDE

$$L[u] = f_2$$

Then, for any any constants c_1 and c_2, $c_1u_1 + c_2u_2$ is a solution of

$$L[u] = c_1f_1 + c_2f_2.$$

That is,

$$L[c_1u_1 + c_2u_2] = c_1f_1 + c_2f_2.$$

In particular, when $f_1 = 0$ and $f_2 = 0$, $L[c_1u_1 + c_2u_2] = c_1f_1 + c_2f_2$ implies that if u_1 and u_2 are solutions of the homogeneous linear PDE $L[u] = 0$, then $c_1u_1 + c_2u_2$ will also be a solution of $L[u] = 0$.

EXAMPLE. Observe that $u_1(x, y) = x^3$ is a solution of the linear PDE $u_{xx} - u_y = 6x$, and $u_2(x, y) = y^2$ is a solution of $u_{xx} - u_y = -2y$. Then, using superposition principle, it is easy to verify that $3u_1(x, y) - 4u_2(x, y)$ will be a solution of $u_{xx} - u_y = 18x + 8y$.

REMARK. Note that the principle of superposition is not valid for nonlinear partial differential equations. This failure makes it difficult to form families of new solutions from an original pair of solutions.

EXAMPLE. Consider the nonlinear first order PDE $u_xu_y - u(u_x + u_y) + u^2 = 0$. Note that e^x and e^y are two solutions of this equation. However, $c_1e^x + c_2e^y$ will not be a solution, unless $c_1 = 0$ or $c_2 = 0$.

Solution: Define $D[u] := (u_x - u)(u_y - u)$. For any $u, v \in C^1$, we have

$$D[u + v] = (u_x + v_x - u - v)(u_y + v_y - u - v)$$

$$= D[u] + D[v] + (u_y - u)(v_x - v) + (u_x - u)(v_y - v).$$

The computation shows that $D[u + v] \neq D[u] + D[v]$ in general. Taking $u = c_1e^x$ and $v = c_2e^y$, an easy computation shows that

$$D[c_1e^x + c_2e^y] = D[c_1e^x] + D[c_2e^y] + (-c_1e^x)(-c_2e^y) = c_1c_2e^{x+y}.$$

Thus, $D[c_1e^x + c_2e^y] = 0$ only if $c_1 = 0$ or $c_2 = 0$.

Well-posed Problems

Who listed three requirements that must be met when formulating an initial and /or boundary value problem. A problem for which the PDE and the data lead to a solution is said to be well posed or correctly posed if the following three conditions are satisfies:

Hadamard's conditions for a well-posed problem are:

1. The solution must exist.

2. The solution should be unique.

3. The solution should depend continuously on the initial and/or boundary data.

If it fails to meet these requirements, it is incorrectly posed.

The conditions $(L[u] = f)$-$(L[c_1u_1 + c_2u_2] = c_1f_1 + c_2f_2)$ require that the equation plus the data for the problem must be such that one and only one solution exists. The third condition states that a small variation of the data for the problem should cause small variation in the solution. As data are generally obtained experimentally and may be subject to numerical approximations, we require that the solution be stable under small variations in initial and/or boundary values. That is, we cannot allow large variations to occur in the solution if the data are altered slightly.

A simple example of an ill posed problem is given below.

EXAMPLE. Consider Cauchy's problem for Laplace's equation in $y \geq 0$:

$$\frac{\partial^2 u}{\partial x^2} + \frac{\partial^2 u}{\partial y^2} = 0,$$
$$u(x,0) = 0,$$
$$u_y(x,0) = \frac{1}{n} \sin nx,$$

where n is a positive integer, is not well-posed.

The solution is given by $u(x,y) = \frac{1}{n^2} \sin(nx) \sin h(ny)$. Now, as $n \to \infty$, $u_y(x, 0) \to 0$ so that for large n the Cauchy data $u(x, 0)$ and $u_y(x, 0)$ can be made arbitrarily small in magnitude. However, the solution $u(x, y)$ oscillates with an amplitude that grows exponentially like e^{ny} as $n \to \infty$. Thus, arbitrarily small data can lead to arbitrarily large variation in solutions and hence the solution is unstable. This violates the condition i.e., the continuous dependence of the solution on the data.

Boundary value problems are not well posed for hyperbolic and parabolic equations. This follows because these are, in general, equations whose solutions evolve in time and their behavior at later times is predicted by their previous states.

EXAMPLE. Consider the hyperbolic equation

$u_{xy} = 0 \ in \ 0 < x < 1, 0 < y < 1$

with the boundary conditions

$u(x, 0) = f_1(x), u(x, 1) = f_2(x) \ for \ 0 \leq x \leq 1,$

$u(0, y) = g_1(y), u(1, y) = g_2(y) \ for \ 0 \leq y \leq 1.$

We shall show that this problem has no solution if the data are prescribed arbitrarily.

Since $u_{xy} = 0$ implies that $u_x(x, y) = $ constant, we have

$u_x(x, 0) = u_x(x, 1).$

In view of the given BC, we have

$$u_x(x,0) = f_1'(x) \quad and \quad u_x(x,1) = f_2'(x)$$

Thus, unless $f_1(x)$ and $f_2(x)$ are prescribed such that $f_1'(x) = f_2'(x)$, the BVP cannot be solved. Therefore, it is incorrectly posed.

Method of Factorization

There is no general methods are available for obtaining the general solutions of second- order PDEs. Sometimes PDE of second-order can be factorized into two first-order equations. The equations

$$u_{\xi\eta} = 0,$$

$$yu_{xx} + (x+y)u_{xy} + xu_{yy} = 0$$

are examples of such equation. It is often much easier to factorize an equation when in its canonical form. But, we can often factorize equations with constant coefficients directly. The method of factorization can be a useful method of solution for hyperbolic and parabolic equations.

EXAMPLE. The equation

$$u_{xx} - u_{yy} + 4(u_x + u) = 0$$

can be written as

$$\left(\frac{\partial}{\partial x} + \frac{\partial}{\partial y} + 2\right)\left(\frac{\partial}{\partial x} - \frac{\partial}{\partial y} + 2\right)u = 0$$

It is equivalent to the pair of first order equations

$u_x - u_y + 2u = v,$

and

$v_x + v_y + 2v = 0.$

EXAMPLE.The hyperbolic equation

$acu_{xy} + au_x + cu_y + u = 0$

can be written as

$$\left(a\frac{\partial}{\partial x} + 1\right)\left(c\frac{\partial}{\partial y} + 1\right)u = 0$$

It is equivalent to

$cu_y + u = v,$

$av_x + v = 0.$

Note: Unlike the case when the coefficients are constant, the differential operators need not commute.

References

- Pinchover, Yehuda; Rubinstein, Jacob (2005). An Introduction to Partial Differential Equations. Cambridge: Cambridge University Press. ISBN 978-0-521-84886-2

- Crank, J.; Nicolson, P. (1947), "A Practical Method for Numerical Evaluation of Solutions of Partial Differential Equations of the Heat-Conduction Type", Proceedings of the Cambridge Philosophical Society, 43: 50–67, Bibcode:1947PCPS...43...50C, doi:10.1017/S0305004100023197

- Cole, K.D.; Beck, J.V.; Haji-Sheikh, A.; Litkouhi, B. (2011), Heat Conduction Using Green's Functions (2nd ed.), CRC Press, ISBN 978-1-43-981354-6

- Perona, P; Malik, J. (1990), "Scale-Space and Edge Detection Using Anisotropic Diffusion", IEEE Transactions on Pattern Analysis and Machine Intelligence, 12 (7): 629–639, doi:10.1109/34.56205

- Mechanical Engineering Design, By Joseph Edward Shigley, Charles R. Mischke, Richard Gordon Budynas, Published 2004 McGraw-Hill Professional, p. 192 ISBN 0-07-252036-1

- Unsworth, J.; Duarte, F. J. (1979), "Heat diffusion in a solid sphere and Fourier Theory", Am. J. Phys., 47 (11): 891–893, Bibcode:1979AmJPh..47..981U, doi:10.1119/1.11601

Linear Differential Equations: An Overview

The solutions that can be summed up together in particular linear combinations to form further solutions are known as linear differential equations. They can be ordinary differential equations and partial differential equations. The chapter closely examines the key concepts of linear differential equations to provide an extensive understanding of the subject.

Linear Differential Equation

In mathematics, linear differential equations are differential equations having solutions which can be added together in particular linear combinations to form further solutions. They equate 0 to a polynomial that is linear in the value and various derivatives of a variable; its linearity means that each term in the polynomial has degree either 0 or 1.

Linear differential equations can be ordinary (ODEs) or partial (PDEs).

The solutions to (homogeneous) linear differential equations form a vector space (unlike non-linear differential equations).

Basic Features

Linear differential equations are of the form

$$Ly = f$$

where the differential operator L is a linear operator, y is the unknown function, and the right hand side f is a given function (called the source term) of the same variable. For a function dependent on time we may write the equation more expressly as

$$Ly(t) = f(t)$$

and, even more precisely by bracketing

$$L[y(t)] = f(t).$$

The linear operator L may be considered to be of the form

$$L_n(y) \equiv \frac{d^n y}{dt^n} + A_1(t)\frac{d^{n-1}y}{dt^{n-1}} + \cdots + A_{n-1}(t)\frac{dy}{dt} + A_n(t)y$$

The linearity condition on L rules out operations such as taking the square of the derivative of y;

but permits, for example, taking the second derivative of y. It is convenient to rewrite this equation in an operator form

$$L_n(y) \equiv \left[D^n + A_1(t)D^{n-1} + \cdots + A_{n-1}(t)D + A_n(t) \right] y$$

where D is the differential operator d/dt (i.e. $Dy = y' = dy/dt$, $D^2y = y'' = d^2y/dt^2$,...), and the A_n are given functions.

Such an equation is said to have order n, the index of the highest derivative of y that is involved.

A typical simple example is the linear differential equation used to model radioactive decay. Let $N(t)$ denote the number of radioactive atoms remaining in some sample of material at time t. Then for some constant $k > 0$, the rate at which the radioactive atoms decay can be modelled by

$$\frac{dN}{dt} = -kN$$

If y is assumed to be a function of only one variable, one speaks about an ordinary differential equation, else the derivatives and their coefficients must be understood as (contracted) vectors, matrices or tensors of higher rank, and we have a (linear) partial differential equation.

The case where f = 0 is called a homogeneous equation and its solutions are called complementary functions. It is particularly important to the solution of the general case, since any complementary function can be added to a solution of the inhomogeneous equation to give another solution (by a method traditionally called *particular integral and complementary function*). When the A_i are numbers, the equation is said to have *constant coefficients*.

Homogeneous Equations with Constant Coefficients

The first method of solving linear homogeneous ordinary differential equations with constant co-efficients is due to Euler, who realized that solutions have the form e^{zx}, for possibly-complex values of z. The exponential function is one of the few functions to keep its shape after differentiation, allowing the sum of its multiple derivatives to cancel out to zero, as required by the equation. Thus, for constant values $A_1,..., A_n$, to solve:

$$y^{(n)} + A_1 y^{(n-1)} + \cdots + A_n y = 0,$$

we set $y = e^{zx}$, leading to

$$z^n e^{zx} + A_1 z^{n-1} e^{zx} + \cdots + A_n e^{zx} = 0.$$

Division by e^{zx} gives the nth-order polynomial:

$$F(z) = z^n + A_1 z^{n-1} + \cdots + A_n = 0.$$

This algebraic equation $F(z) = 0$ is the characteristic equation considered later by Gaspard Monge and Augustin-Louis Cauchy.

Formally, the terms $y^{(k)}$ $(k = 1, 2, \ldots, n)$ of the original differential equation are replaced by z^k. Solving the polynomial gives n values of z, z_1, $\ldots$, z_n. Substitution of any of those values for z into e^{zx} gives a solution $e^{z_i x}$. Since homogeneous linear differential equations obey the superposition principle, any linear combination of these functions also satisfies the differential equation.

When these roots are all distinct, we have n distinct solutions to the differential equation. It can be shown that these are linearly independent, by applying the Vandermonde determinant, and together they form a basis of the space of all solutions of the differential equation.

Examples

$$y''' - 2y''' + 2y'' - 2y' + y = 0$$

has the characteristic equation

$$z^4 - 2z^3 + 2z^2 - 2z + 1 = 0.$$

This has zeroes, i, $-i$, and 1 (multiplicity 2). The solution basis is then

$$e^{ix}, e^{-ix}, e^x, xe^x.$$

This corresponds to the real-valued solution basis

$$\cos x, \sin x, e^x, xe^x.$$

The preceding gave a solution for the case when all zeros are distinct, that is, each has multiplicity 1. For the general case, if z is a (possibly complex) zero (or root) of $F(z)$ having multiplicity m, then, for $k \in \{0, 1, \ldots, m-1\}$, $y = x^k e^{zx}$ is a solution of the ordinary differential equation. Applying this to all roots gives a collection of n distinct and linearly independent functions, where n is the degree of $F(z)$. As before, these functions make up a basis of the solution space.

If the coefficients A_i of the differential equation are real, then real-valued solutions are generally preferable. Since non-real roots z then come in conjugate pairs, so do their corresponding basis functions $x^k e^{zx}$, and the desired result is obtained by replacing each pair with their real-valued linear combinations $\mathrm{Re}(y)$ and $\mathrm{Im}(y)$, where y is one of the pair.

A case that involves complex roots can be solved with the aid of Euler's formula.

Second-order Case

In the $n=2$ case

$$y'' + ay' + by = 0,$$

the characteristic equation is of the form

$$z^2 + az + b = 0.$$

We then can solve for z. There are three particular cases of interest:

- Case #1: Two distinct roots, z_1 and z_2
- Case #2: One real repeated root, z
- Case #3: Complex roots, $\alpha \pm \beta i$

In case #1, the general solution is given by

$$y = c_1 e^{z_1 x} + c_2 e^{z_2 x}.$$

In case #2, the general solution is given by

$$y = (c_1 + c_2 x)e^{zx}.$$

In case #3, the general solution is given, using Euler's equation, by

$$y = c_1 e^{\alpha x} \cos(\beta x) + c_2 e^{\alpha x} \sin(\beta x),$$

$$\alpha = \text{Re}(z),$$

$$\beta = \text{Im}(z).$$

In each case, the constants c_1, c_2 are functions of the initial conditions $y(0), y'(0)$. They can be found by using the values of the initial conditions in the solution equation for y and in the resulting equation for y', giving two equations in the two unknown parameters.

Examples

Given $y'' - 4y' + 5y = 0$. The characteristic equation is $z^2 - 4z + 5 = 0$ which has roots "$2 \pm i$". Thus the solution basis $\{y_1, y_2\}$ is $\{e^{(2+i)x}, e^{(2-i)x}\}$. Now y is a solution if and only if $y = c_1 y_1 + c_2 y_2$ for $c_1, c_2 \in C$.

Because the coefficients are real,

- we are likely not interested in the complex solutions
- our basis elements are mutual conjugates

The linear combinations

$$u_1 = \text{Re}(y_1) = \tfrac{1}{2}(y_1 + y_2) = e^{2x} \cos(x),$$

$$u_2 = \text{Im}(y_1) = \tfrac{1}{2i}(y_1 - y_2) = e^{2x} \sin(x),$$

will give us a real basis in $\{u_1, u_2\}$.

Simple Harmonic Oscillator

The second order differential equation

$$D^2 y = -k^2 y,$$

which represents a simple harmonic oscillator, can be restated as

$$(D^2 + k^2)y = 0.$$

The expression in parenthesis can be factored out, yielding

$$((D + ik)(D - ik))y = 0,$$

which has a pair of linearly independent solutions:

$$(D - ik)y = 0$$

$$(D + ik)y = 0.$$

The solutions are, respectively,

$$y_0 = A_0 e^{ikx}$$

and

$$y_1 = A_1 e^{-ikx}.$$

These solutions provide a basis for the two-dimensional solution space of the second order differential equation: meaning that linear combinations of these solutions will also be solutions. In particular, the following solutions can be constructed

$$y_{0'} = \frac{C_0 e^{ikx} + C_0 e^{-ikx}}{2} = C_0 \cos(kx)$$

and

$$y_{1'} = \frac{C_1 e^{ikx} - C_1 e^{-ikx}}{2i} = C_1 \sin(kx).$$

These last two trigonometric solutions are linearly independent, so they can serve as another basis for the solution space, yielding the following general solution:

$$y_H = C_0 \cos(kx) + C_1 \sin(kx).$$

Damped Harmonic Oscillator

Given the equation for the damped harmonic oscillator:

$$\left(D^2 + \frac{b}{m}D + \omega_0^2\right)y = 0,$$

the expression in parentheses can be factored out: first obtain the characteristic equation by replacing D with z. This equation must be satisfied for all y, thus:

$$z^2 + \frac{b}{m}z + \omega_0^2 = 0.$$

Solve using the quadratic formula:

$$z = \tfrac{1}{2}\left(-\frac{b}{m} \pm \sqrt{\frac{b^2}{m^2} - 4\omega_0^2}\right).$$

Use these characteristic roots to factor the left side of the original differential equation:

$$\left(D + \frac{b}{2m} - \sqrt{\frac{b^2}{4m^2} - \omega_0^2}\right)\left(D + \frac{b}{2m} + \sqrt{\frac{b^2}{4m^2} - \omega_0^2}\right)y = 0.$$

This implies a pair of solutions, one corresponding to

$$\left(D + \frac{b}{2m} - \sqrt{\frac{b^2}{4m^2} - \omega_0^2}\right)y = 0$$

$$\left(D + \frac{b}{2m} + \sqrt{\frac{b^2}{4m^2} - \omega_0^2}\right)y = 0$$

The solutions are, respectively,

$$y_0 = A_0 e^{-\omega x + \sqrt{\omega^2 - \omega_0^2}\,x} = A_0 e^{-\omega x} e^{\sqrt{\omega^2 - \omega_0^2}\,x}$$

$$y_1 = A_1 e^{-\omega x - \sqrt{\omega^2 - \omega_0^2}\,x} = A_1 e^{-\omega x} e^{-\sqrt{\omega^2 - \omega_0^2}\,x}$$

where $\omega = b/2m$. From this linearly independent pair of solutions can be constructed another linearly independent pair which thus serve as a basis for the two-dimensional solution space:

$$y_H(A_0, A_1)(x) = \left(A_0 \sinh\left(\sqrt{\omega^2 - \omega_0^2}\,x\right) + A_1 \cosh\left(\sqrt{\omega^2 - \omega_0^2}\,x\right)\right)e^{-\omega x}.$$

However, if $|\omega| < |\omega_0|$ then it is preferable to get rid of the consequential imaginaries, expressing the general solution as

$$y_H(A_0, A_1)(x) = \left(A_0 \sin\left(\sqrt{\omega_0^2 - \omega^2}\,x\right) + A_1 \cos\left(\sqrt{\omega_0^2 - \omega^2}\,x\right)\right)e^{-\omega x}.$$

This latter solution corresponds to the underdamped case, whereas the former one corresponds to the overdamped case: the solutions for the underdamped case oscillate whereas the solutions for the overdamped case do not.

Nonhomogeneous Equation with Constant Coefficients

To obtain the solution to the nonhomogeneous equation (sometimes called inhomogeneous equation), find a particular integral $y_p(x)$ by the method of undetermined coefficients, method of variation of parameters or the use of the exponential response formula (below); the general solution to the linear differential equation is the sum of the general solution of the related homogeneous equation and the particular integral. Or, when the initial conditions are set, use Laplace transform to obtain the particular solution directly.

Suppose we face

$$\frac{d^n y(x)}{dx^n} + A_1 \frac{d^{n-1} y(x)}{dx^{n-1}} + \cdots + A_n y(x) = f(x).$$

For later convenience, define the characteristic polynomial

$$P(z) = z^n + A_1 z^{n-1} + \cdots + A_n.$$

We find a solution basis $\{y_1(x), y_2(x), \ldots, y_n(x)\}$ for the homogeneous ($f(x) = 0$) case. We now seek a particular integral $y_p(x)$ by the variation of parameters method. Let the coefficients of the linear combination be functions of x:

$$y_p(x) = u_1(x) y_1(x) + u_2(x) y_2(x) + \cdots + u_n(x) y_n(x).$$

For ease of notation we will drop the dependency on x (i.e. the various (x)). Using the operator notation $D = d/dx$, the ODE in question is $P(D)y = f$; so

$$f = P(D)y_p = P(D)(u_1 y_1) + P(D)(u_2 y_2) + \cdots + P(D)(u_n y_n).$$

With the constraints

$$0 = u_1' y_1 + u_2' y_2 + \cdots + u_n' y_n$$
$$0 = u_1' y_1' + u_2' y_2' + \cdots + u_n' y_n'$$
$$\cdots$$
$$0 = u_1' y_1^{(n-2)} + u_2' y_2^{(n-2)} + \cdots + u_n' y_n^{(n-2)}$$

the parameters commute out,

$$f = u_1 P(D) y_1 + u_2 P(D) y_2 + \cdots + u_n P(D) y_n + u_1' y_1^{(n-1)} + u_2' y_2^{(n-1)} + \cdots + u_n' y_n^{(n-1)}.$$

But $P(D)y_j = 0$, therefore

$$f = u_1' y_1^{(n-1)} + u_2' y_2^{(n-1)} + \cdots + u_n' y_n^{(n-1)}.$$

This, with the constraints, gives a linear system in the u'_j. This much can always be solved; in fact, combining Cramer's rule with the Wronskian,

$$u_j' = (-1)^{n+j} \frac{W(y_1, \ldots, y_{j-1}, y_{j+1} \ldots, y_n)\binom{0}{f}}{W(y_1, y_2, \ldots, y_n)}.$$

In the very non-standard notation used above, one should take the i,n-minor of W and multiply it by f. That's why we get a minus-sign. Alternatively, forget about the minus sign and just compute the determinant of the matrix obtained by substituting the j-th W column with (0, 0, ..., f).

The rest is a matter of integrating u'_j.

The particular integral is not unique; $y_p + c_1 y_1 + \cdots + c_n y_n$ also satisfies the ODE for any set of constants c_j.

Exponential Response Formula

The particular solution of

$$P(D)y = \sum_i \left(a_i e^{r_i x} \right)$$

can be found as

$$y_p = \sum_i \left(\frac{a_i}{P(r_i)} e^{r_i x} \right).$$

Example

Suppose $y'' - 4y' + 5y = \sin(kx)$.

We take the solution basis found above $\{e^{(2+i)x} = y_1(x), e^{(2-i)x} = y_2(x)\}$.

$$W = \begin{vmatrix} e^{(2+i)x} & e^{(2-i)x} \\ (2+i)e^{(2+i)x} & (2-i)e^{(2-i)x} \end{vmatrix} = e^{4x} \begin{vmatrix} 1 & 1 \\ 2+i & 2-i \end{vmatrix} = -2ie^{4x}$$

$$u_1' = \frac{1}{W} \begin{vmatrix} 0 & e^{(2-i)x} \\ \sin(kx) & (2-i)e^{(2-i)x} \end{vmatrix} = -\tfrac{i}{2} \sin(kx) e^{(-2-i)x}$$

$$u_2' = \frac{1}{W} \begin{vmatrix} e^{(2+i)x} & 0 \\ (2+i)e^{(2+i)x} & \sin(kx) \end{vmatrix} = \tfrac{i}{2} \sin(kx) e^{(-2+i)x}.$$

Using the list of integrals of exponential functions

$$u_1 = -\frac{i}{2}\int \sin(kx)e^{(-2-i)x}\,dx = \frac{ie^{(-2-i)x}}{2(3+4i+k^2)}\big((2+i)\sin(kx)+k\cos(kx)\big)$$

$$u_2 = \frac{i}{2}\int \sin(kx)e^{(-2+i)x}\,dx = \frac{ie^{(i-2)x}}{2(3-4i+k^2)}\big((i-2)\sin(kx)-k\cos(kx)\big).$$

and so

$$y_p = u_1(x)y_1(x)+u_2(x)y_2(x)$$

$$=\frac{i}{2(3+4i+k^2)}\big((2+i)\sin(kx)+k\cos(kx)\big)+\frac{i}{2(3-4i+k^2)}\big((i-2)\sin(kx)-k\cos(kx)\big)$$

$$=\frac{(5-k^2)\sin(kx)+4k\cos(kx)}{(3+k^2)^2+16}.$$

(Notice that u_1 and u_2 had factors that canceled y_1 and y_2; that is typical.)

For interest's sake, this ODE has a physical interpretation as a driven damped harmonic oscillator; y_p represents the steady state, and $c_1 y_1 + c_2 y_2$ is the transient.

As $\sin(kx)=\dfrac{e^{ikx}-e^{-ikx}}{2i}$, the method of the exponential response formula produces

$$y_p = \frac{\dfrac{e^{ikx}}{5-k^2-4ik}-\dfrac{e^{-ikx}}{5-k^2+4ik}}{2i}$$

$$=\frac{(5-k^2)\sin(kx)+4k\cos(kx)}{(5-k^2)^2+(4k)^2},$$

the same answer as above.

Equation with Variable Coefficients

A linear ODE of order n with variable coefficients has the general form

$$p_n(x)y^{(n)}(x)+p_{n-1}(x)y^{(n-1)}(x)+\cdots+p_0(x)y(x)=r(x).$$

Examples

A simple example is the Cauchy–Euler equation often used in engineering

$$x^n y^{(n)}(x)+a_{n-1}x^{n-1}y^{(n-1)}(x)+\cdots+a_0 y(x)=0.$$

First-order Equation with Variable Coefficients

Examples

Solve the equation

$$y'(x) + 3y(x) = 2$$

with the initial condition

$$y(0) = 2.$$

Using the general solution method:

$$y = e^{-3x}\left(\int 2e^{3x}\,dx + \kappa\right).$$

The indefinite integral is solved to give:

$$y = e^{-3x}\left((2/3)e^{3x} + \kappa\right).$$

Then we can reduce to:

$$y = 2/3 + \kappa e^{-3x}.$$

where $\kappa = 4/3$ from the initial condition.

A linear ODE of order 1 with variable coefficients has the general form

$$Dy(x) + f(x)y(x) = g(x).$$

Where D is the differential operator. Equations of this form can be solved by multiplying the integrating factor

$$e^{\int f(x)\,dx}$$

throughout to obtain

$$Dy(x)e^{\int f(x)\,dx} + f(x)y(x)e^{\int f(x)\,dx} = g(x)e^{\int f(x)\,dx},$$

which simplifies due to the product rule (applied backwards) to

$$D\left(y(x)e^{\int f(x)\,dx}\right) = g(x)e^{\int f(x)\,dx}$$

which, on integrating both sides and solving for $y(x)$ gives:

$$y(x) = e^{-\int f(x)\,dx}\left(\int g(x)e^{\int f(x)\,dx}\,dx + \kappa\right).$$

In other words: The solution of a first-order linear ODE

$$y'(x) + f(x)y(x) = g(x),$$

with coefficients that may or may not vary with x, is:

$$y = e^{-a(x)}\left(\int g(x)e^{a(x)}\,dx + \kappa\right)$$

where κ is the constant of integration, and

$$a(x) = \int f(x)dx.$$

A compact form of the general solution based on a Green's function is

$$y(x) = \int_a^x [y(a)\delta(t-a) + g(t)]e^{-\int_t^x f(u)du}\,dt.$$

where $\delta(x)$ is the generalized Dirac delta function.

Examples

Consider a first order differential equation with constant coefficients:

$$\frac{dy}{dx} + by = 1.$$

This equation is particularly relevant to first order systems such as RC circuits and mass-damper systems.

In this case, $f(x) = b$, $g(x) = 1$.

Hence its solution is

$$y(x) = e^{-bx}\left(\frac{e^{bx}}{b} + C\right) = \frac{1}{b} + Ce^{-bx}.$$

Systems of Linear Differential Equations

An arbitrary linear ordinary differential equation or even a system of such equations can be converted into a first order system of linear differential equations by adding variables for all but the highest order derivatives. A linear system can be viewed as a single equation with a vector-valued variable. The general treatment is analogous to the treatment above of ordinary first order linear differential equations, but with complications stemming from noncommutativity of matrix multiplication.

To solve

$$\begin{cases} y'(x) & = A(x)y(x) + b(x) \\ y(x_0) & = y_0 \end{cases}$$

(here $y(x)$ is a vector or matrix, and $A(x)$ is a matrix), let $U(x)$ be the solution of $y'(x) = A(x)y(x)$ with $U(x_0) = I$ (the identity matrix). U is a fundamental matrix for the equation — the columns of U form a complete linearly independent set of solutions for the homogeneous equation. After substituting $y(x) = U(x)z(x)$, the equation $y'(x) = A(x)y(x) + b(x)$ simplifies to $U(x)z'(x) = b(x)$. Thus,

$$y(x) = U(x)y_0 + U(x)\int_{x_0}^x U^{-1}(t)b(t)dt$$

If $A(x_1)$ commutes with $A(x_2)$ for all x_1 and x_2, then

$$U(x) = e^{\int_{x_0}^{x} A(x)dx}$$

and thus

$$U^{-1}(x) = e^{-\int_{x_0}^{x} A(x)dx},$$

but in the general case there is no closed form solution, and an approximation method such as Magnus expansion may have to be used. Note that the exponentials are matrix exponentials.

Linear Independence

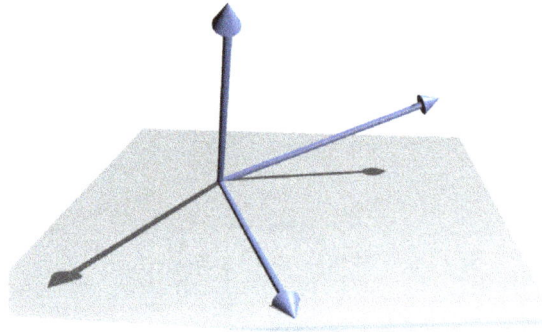

Linearly independent vectors in $\mathbb{R}^3$

Linearly dependent vectors in a plane in $\mathbb{R}^3$.

In the theory of vector spaces, a set of vectors is said to be linearly dependent if one of the vectors in the set can be defined as a linear combination of the others; if no vector in the set can be written in this way, then the vectors are said to be linearly independent. These concepts are central to the definition of dimension.

A vector space can be of finite-dimension or infinite-dimension depending on the number of linearly independent basis vectors. The definition of linear dependence and the ability to determine whether a subset of vectors in a vector space is linearly dependent are central to determining a basis for a vector space.

Definition

The vectors in a subset $S = \{\vec{v}_1, \vec{v}_2, \ldots, \vec{v}_n\}$ of a vector space V are said to be *linearly dependent*, if there exist a *finite* number of *distinct* vectors $\vec{v}_1, \vec{v}_2, \ldots, \vec{v}_k$ in S and scalars $a_1, a_2, \ldots, a_k$, not all zero, such that

$$a_1\vec{v}_1 + a_2\vec{v}_2 + \cdots + a_k\vec{v}_k = \vec{0},$$

where $\vec{0}$ denotes the zero vector.

Notice that if not all of the scalars are zero, then at least one is non-zero, say a_1, in which case this equation can be written in the form

$$\vec{v}_1 = \frac{-a_2}{a_1}\vec{v}_2 + \cdots + \frac{-a_k}{a_1}\vec{v}_k.$$

Thus, v_1 is shown to be a linear combination of the remaining vectors.

The vectors in a set $T = \{\vec{v}_1, \vec{v}_2, \ldots, \vec{v}_n\}$ are said to be *linearly independent* if the equation

$$a_1\vec{v}_1 + a_2\vec{v}_2 + \cdots + a_n\vec{v}_n = \vec{0},$$

can only be satisfied by $a_i = 0$ for $i = 1, \ldots, n$. This implies that no vector in the set can be represented as a linear combination of the remaining vectors in the set. In other words, a set of vectors is linearly independent if the only representations of $\vec{0}$ as a linear combination of its vectors is the trivial representation in which all the scalars a_i are zero.

Infinite Dimensions

In order to allow the number of linearly independent vectors in a vector space to be countably infinite, it is useful to define linear dependence as follows. More generally, let V be a vector space over a field K, and let $\{v_i \mid i \in I\}$ be a family of elements of V. The family is *linearly dependent* over K if there exists a family $\{a_j \mid j \in J\}$ of elements of K, not all zero, such that

$$\sum_{j \in J} a_j v_j = 0$$

where the index set J is a nonempty, finite subset of I.

A set X of elements of V is *linearly independent* if the corresponding family $\{x\}_{x \in X}$ is linearly independent. Equivalently, a family is dependent if a member is in the linear span of the rest of the family, i.e., a member is a linear combination of the rest of the family. The trivial case of the empty family must be regarded as linearly independent for theorems to apply.

A set of vectors which is linearly independent and spans some vector space, forms a basis for that vector space. For example, the vector space of all polynomials in x over the reals has the (infinite) subset $\{1, x, x^2, \ldots\}$ as a basis.

Geometric Meaning

A geographic example may help to clarify the concept of linear independence. A person describing the location of a certain place might say, "It is 3 miles north and 4 miles east of here." This is sufficient information to describe the location, because the geographic coordinate system may be considered as a 2-dimensional vector space (ignoring altitude and the curvature of the Earth's surface). The person might add, "The place is 5 miles northeast of here." Although this last statement is *true*, it is not necessary.

In this example the "3 miles north" vector and the "4 miles east" vector are linearly independent. That is to say, the north vector cannot be described in terms of the east vector, and vice versa. The third "5 miles northeast" vector is a linear combination of the other two vectors, and it makes the set of vectors *linearly dependent*, that is, one of the three vectors is unnecessary.

Also note that if altitude is not ignored, it becomes necessary to add a third vector to the linearly independent set. In general, n linearly independent vectors are required to describe any location in n-dimensional space.

Evaluating Linear Independence

Vectors in R²

Three vectors: Consider the set of vectors $v_1 = (1, 1)$, $v_2 = (-3, 2)$ and $v_3 = (2, 4)$, then the condition for linear dependence seeks a set of non-zero scalars, such that

$$a_1 \begin{Bmatrix} 1 \\ 1 \end{Bmatrix} + a_2 \begin{Bmatrix} -3 \\ 2 \end{Bmatrix} + a_3 \begin{Bmatrix} 2 \\ 4 \end{Bmatrix} = \begin{Bmatrix} 0 \\ 0 \end{Bmatrix},$$

or

$$\begin{bmatrix} 1 & -3 & 2 \\ 1 & 2 & 4 \end{bmatrix} \begin{Bmatrix} a_1 \\ a_2 \\ a_3 \end{Bmatrix} = \begin{Bmatrix} 0 \\ 0 \end{Bmatrix}.$$

Row reduce this matrix equation by subtracting the first row from the second to obtain,

$$\begin{bmatrix} 1 & -3 & 2 \\ 0 & 5 & 2 \end{bmatrix} \begin{Bmatrix} a_1 \\ a_2 \\ a_3 \end{Bmatrix} = \begin{Bmatrix} 0 \\ 0 \end{Bmatrix}.$$

Continue the row reduction by (i) dividing the second row by 5, and then (ii) multiplying by 3 and adding to the first row, that is

$$\begin{bmatrix} 1 & 0 & 16/5 \\ 0 & 1 & 2/5 \end{bmatrix} \begin{Bmatrix} a_1 \\ a_2 \\ a_3 \end{Bmatrix} = \begin{Bmatrix} 0 \\ 0 \end{Bmatrix}.$$

We can now rearrange this equation to obtain

$$\begin{bmatrix} 1 & 0 \\ 0 & 1 \end{bmatrix}\begin{Bmatrix} a_1 \\ a_2 \end{Bmatrix} = \begin{Bmatrix} a_1 \\ a_2 \end{Bmatrix} = -a_3 \begin{Bmatrix} 16/5 \\ 2/5 \end{Bmatrix}.$$

which shows that non-zero a_i exist $v_3 = (2, 4)$ can be defined in terms of $v_1 = (1, 1)$, $v_2 = (-3, 2)$. Thus, the three vectors are linearly dependent.

Two vectors: Now consider the linear dependence of the two vectors $v_1 = (1, 1)$, $v_2 = (-3, 2)$, and check,

$$a_1\begin{Bmatrix} 1 \\ 1 \end{Bmatrix} + a_2\begin{Bmatrix} -3 \\ 2 \end{Bmatrix} = \begin{Bmatrix} 0 \\ 0 \end{Bmatrix},$$

or

$$\begin{bmatrix} 1 & -3 \\ 1 & 2 \end{bmatrix}\begin{Bmatrix} a_1 \\ a_2 \end{Bmatrix} = \begin{Bmatrix} 0 \\ 0 \end{Bmatrix}.$$

The same row reduction presented above yields,

$$\begin{bmatrix} 1 & 0 \\ 0 & 1 \end{bmatrix}\begin{Bmatrix} a_1 \\ a_2 \end{Bmatrix} = \begin{Bmatrix} 0 \\ 0 \end{Bmatrix}.$$

This shows that $a_i = 0$, which means that the vectors $v_1 = (1, 1)$ and $v_2 = (-3, 2)$ are linearly independent.

Vectors in R⁴

In order to determine if the three vectors in R⁴,

$$v_1 = \begin{Bmatrix} 1 \\ 4 \\ 2 \\ -3 \end{Bmatrix}, v_2 = \begin{Bmatrix} 7 \\ 10 \\ -4 \\ -1 \end{Bmatrix}, v_3 = \begin{Bmatrix} -2 \\ 1 \\ 5 \\ -4 \end{Bmatrix}.$$

are linearly dependent, form the matrix equation,

$$\begin{bmatrix} 1 & 7 & -2 \\ 4 & 10 & 1 \\ 2 & -4 & 5 \\ -3 & -1 & -4 \end{bmatrix}\begin{Bmatrix} a_1 \\ a_2 \\ a_3 \end{Bmatrix} = \begin{Bmatrix} 0 \\ 0 \\ 0 \\ 0 \end{Bmatrix}.$$

Row reduce this equation to obtain,

$$\begin{bmatrix} 1 & 7 & -2 \\ 0 & -18 & 9 \\ 0 & 0 & 0 \\ 0 & 0 & 0 \end{bmatrix} \begin{Bmatrix} a_1 \\ a_2 \\ a_3 \end{Bmatrix} = \begin{Bmatrix} 0 \\ 0 \\ 0 \\ 0 \end{Bmatrix}.$$

Rearrange to solve for v_3 and obtain,

$$\begin{bmatrix} 1 & 7 \\ 0 & -18 \end{bmatrix} \begin{Bmatrix} a_1 \\ a_2 \end{Bmatrix} = -a_3 \begin{Bmatrix} -2 \\ 9 \end{Bmatrix}.$$

This equation is easily solved to define non-zero a_i,

$$a_1 = -3a_3/2, a_2 = a_3/2,$$

where a_3 can be chosen arbitrarily. Thus, the vectors v_1, v_2 and v_3 are linearly dependent.

Alternative Method using Determinants

An alternative method of fact that n vectors in $\mathbb{R}^n$ are linearly independent if and only if the determinant of the matrix formed by taking the vectors as its columns is non-zero.

In this case, the matrix formed by the vectors is

$$A = \begin{bmatrix} 1 & -3 \\ 1 & 2 \end{bmatrix}.$$

We may write a linear combination of the columns as

$$A\Lambda = \begin{bmatrix} 1 & -3 \\ 1 & 2 \end{bmatrix} \begin{bmatrix} \lambda_1 \\ \lambda_2 \end{bmatrix}.$$

We are interested in whether $A\Lambda = \mathbf{o}$ for some nonzero vector Λ. This depends on the determinant of A, which is

$$\det A = 1 \cdot 2 - 1 \cdot (-3) = 5 \neq 0.$$

Since the determinant is non-zero, the vectors $(1, 1)$ and $(-3, 2)$ are linearly independent.

Otherwise, suppose we have m vectors of n coordinates, with $m < n$. Then A is an $n \times m$ matrix and Λ is a column vector with m entries, and we are again interested in $A\Lambda = \mathbf{o}$. As we saw previously, this is equivalent to a list of n equations. Consider the first m rows of A, the first m equations; any solution of the full list of equations must also be true of the reduced list. In fact, if $\langle i_1,...,i_m \rangle$ is any list of m rows, then the equation must be true for those rows.

$$A_{\langle i_1,...,i_m \rangle}\Lambda = 0.$$

Furthermore, the reverse is true. That is, we can test whether the m vectors are linearly dependent by testing whether

$$\det A_{\langle i_1,\ldots,i_m\rangle} = 0$$

for all possible lists of m rows. (In case $m = n$, this requires only one determinant, as above. If $m > n$, then it is a theorem that the vectors must be linearly dependent.) This fact is valuable for theory; in practical calculations more efficient methods are available.

Natural basis vectors

Let $V = R^n$ and consider the following elements in V, known as the natural basis vectors:

$$
\begin{aligned}
e_1 &= (1,0,0,\ldots,0)\\
e_2 &= (0,1,0,\ldots,0)\\
&\vdots\\
e_n &= (0,0,0,\ldots,1).
\end{aligned}
$$

Then e_1, e_2, ..., e_n are linearly independent.

Proof

Suppose that a_1, a_2, ..., a_n are elements of R such that

$$a_1 e_1 + a_2 e_2 + \cdots + a_n e_n = 0.$$

Since

$$a_1 e_1 + a_2 e_2 + \cdots + a_n e_n = (a_1, a_2, \ldots, a_n),$$

then $a_i = 0$ for all i in $\{1, \ldots, n\}$.

Linear Independence of Basis Functions

Let V be the vector space of all functions of a real variable t. Then the functions e^t and e^{2t} in V are linearly independent.

Proof

Suppose a and b are two real numbers such that

$$ae^t + be^{2t} = 0$$

for *all* values of t. We need to show that $a = 0$ and $b = 0$. In order to do this, we divide through by e^t (which is never zero) and subtract to obtain

$$be^t = -a.$$

In other words, the function be^t must be independent of t, which only occurs when $b = 0$. It follows that a is also zero.

Projective Space of linear Dependences

A linear dependence among vectors $v_1, ..., v_n$ is a tuple $(a_1, ..., a_n)$ with n scalar components, not all zero, such that

$$a_1 v_1 + \cdots + a_n v_n = 0.$$

If such a linear dependence exists, then the n vectors are linearly dependent. It makes sense to identify two linear dependences if one arises as a non-zero multiple of the other, because in this case the two describe the same linear relationship among the vectors. Under this identification, the set of all linear dependences among $v_1, ..., v_n$ is a projective space.

Linear Dependence between Random Variables

The covariance is sometimes called a measure of "linear dependence" between two random variables. That does not mean the same thing as in the context of linear algebra. When the covariance is normalized, one obtains the correlation matrix. From it, one can obtain the Pearson coefficient, which gives the goodness of the fit for the best possible linear function describing the relation between the variables. In this sense covariance is a linear gauge of dependence.

Wronskian

In mathematics, the Wronskian (or Wrońskian) is a determinant introduced by Józef Hoene-Wroński (1776) and named by Thomas Muir (1882, Chapter XVIII). It is used in the study of differential equations, where it can sometimes show linear independence in a set of solutions.

Definition

The Wronskian of two differentiable functions f and g is $W(f,g) = fg' - gf'$.

More generally, for n real- or complex-valued functions $f_1, \ldots, f_n$, which are $n-1$ times differentiable on an interval I, the Wronskian $W(f_1, \ldots, f_n)$ as a function on I is defined by

$$W(f_1,...,f_n)(x) = \begin{vmatrix} f_1(x) & f_2(x) & \cdots & f_n(x) \\ f_1'(x) & f_2'(x) & \cdots & f_n'(x) \\ \vdots & \vdots & \ddots & \vdots \\ f_1^{(n-1)}(x) & f_2^{(n-1)}(x) & \cdots & f_n^{(n-1)}(x) \end{vmatrix}, \quad x \in I.$$

That is, it is the determinant of the matrix constructed by placing the functions in the first row, the first derivative of each function in the second row, and so on through the $(n-1)$th derivative, thus forming a square matrix sometimes called a fundamental matrix.

When the functions f_i are solutions of a linear differential equation, the Wronskian can be found explicitly using Abel's identity, even if the functions f_i are not known explicitly.

The Wronskian and Linear Independence

If the functions f_i are linearly dependent, then so are the columns of the Wronskian as differentiation is a linear operation, so the Wronskian vanishes. Thus, the Wronskian can be used to show that a set of differentiable functions is linearly independent on an interval by showing that it does not vanish identically. It may, however, vanish at isolated points.

A common misconception is that $W = 0$ everywhere implies linear dependence, but Peano (1889) pointed out that the functions x^2 and $|x|x$ have continuous derivatives and their Wronskian vanishes everywhere, yet they are not linearly dependent in any neighborhood of 0. There are several extra conditions which ensure that the vanishing of the Wronskian in an interval implies linear dependence. Peano (1889) observed that if the functions are analytic, then the vanishing of the Wronskian in an interval implies that they are linearly dependent. (Peano published his example twice, because the first time he published it an editor Paul Mansion, who had written a textbook incorrectly claiming that the vanishing of the Wronskian implies linear dependence, added a footnote to Peano's paper claiming that this result is correct as long as neither function is identically zero. Peano's second paper pointed out that this footnote was nonsense.) Bocher (1901) gave several other conditions for the vanishing of the Wronskian to imply linear dependence; for example, if the Wronskian of n functions is identically zero and the n Wronskians of $n - 1$ of them do not all vanish at any point then the functions are linearly dependent. Wolsson (1989a) gave a more general condition that together with the vanishing of the Wronskian implies linear dependence.

Over fields of positive characteristic p the Wronskian may vanish even for linearly independent polynomials; for example, the Wronskian of x^p and 1 is identically 0.

Application to Linear Differential Equations

In general, for an n th order linear differential equation, if $(n-1)$ solutions are known, the last one can be determined by using the Wronskian.

Consider the second order differential equation in Lagrange's notation

$$y'' = ay' + by$$

where $a(x), b(x)$ are known functions of x and $y(x)$ is the yet to be determined function. Let us call y_1, y_2 the two solutions of the equation and form their Wronskian

$$W(x) = y_1 y_2' - y_2 y_1'$$

Then differentiating $W(x)$ and using the fact that y_i obey the above differential equation shows that

$$W'(x) = aW(x)$$

Therefore, the Wronskian obeys a simple first order differential equation and can be exactly solved:

$$W(x) = e^{A(x)} \text{ where } A'(x) = a(x)$$

Now suppose that we know one of the solutions, say y_2. Then, by the definition of the Wronskian,

y_1 obeys a first order differential equation:

$$y_1' - \frac{y_2'}{y_2} y_1 = -W(x)/y_2$$

and can be solved exactly (at least in theory).

The method is easily generalized to higher order equations.

Generalized Wronskians

For n functions of several variables, a generalized Wronskian is a determinant of an n by n matrix with entries $D_i(f_j)$ (with $0 \le i < n$), where each D_i is some constant coefficient linear partial differential operator of order i. If the functions are linearly dependent then all generalized Wronskians vanish. As in the 1 variable case the converse is not true in general: if all generalized Wronskians vanish, this does not imply that the functions are linearly dependent. However, the converse is true in many special cases. For example, if the functions are polynomials and all generalized Wronskians vanish, then the functions are linearly dependent. Roth used this result about generalized Wronskians in his proof of Roth's theorem.

Basic Theory for Linear Equations

In this section the meaning that is attached to a general solution of the differential equation and some of its properties are studied. We stick our attention to second order equations to start with and extend the study for an n-th order linear equation. The extension is not hard at all. As usual let I ⊆ R be an interval. Consider

$$a_0(t)x''(t) + a_1(t)x'(t) + a_2(t)x(t) = 0, \quad a_0(t) \neq 0, \ t \in I.$$
$$(1)$$

Later we shall study structure of solutions of a non-homogeneous equation of second order. Let us define an operator L on the space of twice differentiable functions defined on I by the following relation

$$L(y)(t) = a_0(t)y''(t) + a_1(t)y'(t) + a_2(t)y(t) \quad \text{and} \quad a_0(t) \neq 0, \ t \in I.$$
$$(2)$$

With L in hand, (1) is

$$L(x) = 0 \text{ on } I.$$

The linearity of the differential operator tell us that:

Lemma. The operator L is linear on the space of twice differential functions on I.

Proof: Let y_1 and y_2 be any two twice differentiable functions on I. Let c_1 and c_2 be any constants. For the linearity of L we need to show

$$L(c_1 y_1 + c_2 y_2) = c_1 L(y_1) + c_2 L(y_2) \text{ on } I$$

which is a simple consequence of the linearity of the differential operator.

As an immediate consequence of the Lemma, we have the superposition principle:

Theorem. (Super Position Principle) Suppose x_1 and x_2 satisfy the equation (1) for $t \in I$. Then,

$$c_1 x_1 + c_2 x_2,$$

also satisfies (1), where c_1 and c_2 are any constants.

The proof is easy and hence, omitted. The first of the following examples illustrates Theorem 2.3.2 while the second one shows that the linearity cannot be dropped.

Example. (i) Consider the differential equation for the linear harmonic oscillator, namely

$$x'' + \lambda^2 x = 0, \ \lambda \in \mathbb{R}.$$

Both $\sin \lambda x$ and $\cos \lambda x$ are two solutions of this equation and

$$c_1 \sin \lambda x + c_2 \cos \lambda x,$$

is also a solution, where c_1 and c_2 are constants.

(ii) The differential equation

$$x'' = -x'^2,$$

admits two solutions

$$x_1(t) = \log(t + a_1) + a_2 \text{ and } x_2(t) = \log(t + a_1),$$

where a_1 and a_2 are constants. With the values of $c_1 = 3$ and $c_2 = -1$,

$$x(t) = c_1 x_1(t) + c_2 x_2(t),$$

does not satisfy the given equation. We note that the given equation is nonlinear.

Lemma and Theorem which prove the principle of superposition for the linear equations of second order have a natural extension to linear equations of order $n(n > 2)$. Let

Let
$$L(y) = a_0(t)y^{(n)} + a_1(t)y^{(n-1)} + \cdots + a_n(t)y, \quad t \in I \qquad (3)$$

where $a_0 t \neq 0$ on I. The general n-th order linear differential equation may be written as

$$L(x) = 0, \qquad (4)$$

where L is the operator defined by the relation (3). As a consequence of the definition, we have:

Lemma The operator L defined by (3), is a linear operator on the space of all n times differentiable functions defined on I.

Theorem. Suppose x_1, x_2,..., x_n satisfy the equation (4). Then,

$$c_1 x_1 + c_2 x_2 + \cdots + c_n x_n,$$

also satisfies (4), where c_1, c_2,...,c_n are arbitrary constants.

The proofs of the Lemma and Theorem are easy and hence omitted.

Theorem allows us to define a general solution of (4) given an additional hypothesis that the set of solutions x_1, x_2,...,x_n is linearly independent. Under these assumptions later we actually show that any solution x of (2) is indeed a linear combination of x_1, x_2,...,x_n.

Definition. Let x_1, x_2,...,x_n be n linearly independent solutions of (4). Then,

$$c_1 x_1 + c_2 x_2 + \cdots + c_n x_n,$$

is called the general solution of (4), where c_1, c_2,...,c_n are arbitrary constants.

Example. Consider the equation

$$x'' - \frac{2}{t^2} x = 0, \quad 0 < t < \infty.$$

We note that $x_1(t) = t^2$ and $x_2(t) = \frac{1}{t}$ are 2 linearly independent solutions on $0 < t < \infty$. A general solution x is

$$x(t) = c_1 t^2 + \frac{c_2}{t}, \quad 0 < t < \infty.$$

Example. $x_1(t) = t$, $x_2(t) = t^2$, $x_3(t) = t^3$, t > 0 are three linearly independent solutions of the equation

$$t^3 x''' - 3t^2 x'' + 6tx' - 6x = 0, \quad t > 0.$$

The general solution x is

$$x(t) = c_1 t + c_2 t^2 + c_3 t^3, t > 0.$$

We again recall that Theorems state that the linear combinations of solutions of a linear equation is yet another solution. The question now is whether this property can be used to generate the general solution for a given linear equation. The answer indeed is in affirmative. Here we make use of the interplay between linear independence of solutions and the Wronskian. The following preparatory result is needed for further discussion. We recall the equation (2) for the definition of L.

Lemma. If x_1 and x_2 are linearly independent solutions of the equation L(x) = 0 on I, then the Wronskian of x_1 and x_2, namely, W $[x_1(t), x_2(t)]$ is never zero on I.

Proof: Suppose on the contrary, there exist $t_0 \in I$ at which $W[x_1(t_0), x_2(t_0)] = 0$. Then, the system of linear algebraic equations for c_1 and c_2

$$\left. \begin{array}{l} c_1 x_1(t_0) + c_2(t) x_2(t_0) = 0 \\ c_1 x_1'(t_0) + c_2(t) x_2'(t_0) = 0 \end{array} \right\}, \tag{5}$$

has a non-trivial solution. For such a nontrivial solution (c_1, c_2) of (5), we define

$$x(t) = c_1 x_1(t) + c_2 x_2(t), \quad t \in I.$$

By Theorem, x is a solution of the equation (1) and

$$x(t_0) = 0 \text{ and } x'(t_0) = 0.$$

Since an initial value problem for $L(x) = 0$ admits only one solution, we therefore have $x(t) \equiv 0$, $t \in I$, which means that

$$c_1 x_1(t) + c_2 x_2(t) \equiv 0, \quad t \in I,$$

with at least one of c_1 and c_2 is non-zero or else, x_1, x_2 are linearly dependent on I, which is a contradiction. So the Wronskian $W[x_1, x_2]$ cannot vanish at any point of the interval I.

As a consequence of the above lemma an interesting corollary is:

Corollary. The Wronskian of two solutions of $L(x) = 0$ is either identically zero if the solutions are linearly dependent on I or never zero if the solutions are linearly independent on I.

Lemma has an immediate generalization of to the equations of order $n(n > 2)$. The following lemma is stated without proof.

Lemma. If $x_1(t)$, $x_2(t)$,..., $x_n(t)$ are linearly independent solutions of the equation (4) which exist on I, then the Wronskian

$$W[x_1(t), x_2(t), \cdots, x_n(t)],$$

is never zero on I. The converse also holds.

Example. Consider Examples. The linearly independent solutions of the differential equation in Example are $x_1(t) = t^2$, $x_2(t) = 1/t$. The Wronskian of these solutions is

$$W[x_1(t), x_2(t)] = -3 \neq 0 \text{ for } t \in (-\infty, \infty).$$

The Wronskian of the solutions in Example is given by

$$W[x_1(t), x_2(t), x_3(t)] = 2t^3 \neq 0$$

when $t > 0$.

The conclusion of the Lemma holds if the equation (4) has n linearly independent solutions. A doubt may occur whether such a set of solutions exist or not. In fact, Example removes such a doubt.

Example. Let

$$L(x) = a_0(t)x''' + a_1(t)x'' + a_1(t)x' + a_3(t)x = 0.$$

Now, let $x_1(t)$, $t \in I$ be the unique solution of the IVP

$$L(x) = 0, \ x(a) = 1, \ x'(a) = 0, \ x''(a) = 0;$$

$x_1(t)$, $t \in I$ be the unique solution of the IVP

$$L(x) = 0, \ x(a) = 0, \ x'(a) = 0, \ x''(a) = 1$$

and $x_3(t)$, $t \in I$ be the unique solution of the IVP

$$L(x) = 0, \ x(a) = 0, \ x'(a) = 0, \ x''(a) = 1$$

where $a \in I$. Obviously $x_1(t)$, $x_2(t)$, $x_3(t)$ are linearly independent, since the value of the Wronskian at the point $a \in I$ is non-zero. For

$$W[x_1(a), x_2(a), x_3(a)] = \begin{vmatrix} 1 & 0 & 0 \\ 0 & 1 & 0 \\ 0 & 0 & 1 \end{vmatrix} = 1 \neq 0.$$

An application of the Lemma justifies the assertion. Thus, a set of three linearly independent solution exists for a homogeneous linear equation of the third order.

Now we establish a major result for a homogeneous linear differential equation of order $n \geq 2$ below.

Theorem. Let x_1, x_2,...,x_n be linearly independent solutions of (4) existing on an interval $I \subseteq R$. Then any solution x of (4) existing on I is of the form

$$x(t) = c_1 x_1(t) + c_2 x_2(t) + \cdots + c_n x_n(t), \ t \in I$$

where c_1, c_2,...,c_n are some constants.

Proof: Let x be any solution of L(x) = 0 on I, and a $\in$ I. Let

$$x(a) = a_1, x'(a) = a_2, \cdots, x^{(n-1)} = a_n.$$

Consider the following system of equation:

$$\left. \begin{array}{l} c_1 x_1(a) + c_2 x_2(a) + \cdots + c_n x_n(a) = a_1 \\ c_1 x_1'(a) + c_2 x_2'(a) + \cdots + c_n x_n'(a) = a_2 \\ \cdots\cdots\cdots\cdots\cdots\cdots\cdots\cdots\cdots\cdots\cdots \\ c_1 x_1^{(n-1)}(a) + c_2 x_2^{(n-1)}(a) + \cdots + c_n x_n^{(n-1)}(a) = a_n \end{array} \right\}. \qquad (6)$$

We can solve system of equations (6) for $c_1, c_2,...,c_n$. The determinant of the coefficients of $c_1, c_2,...,c_n$ in the above system is not zero and since the Wronskian of $x_1, x_2,...,x_n$ at the point a is different from zero by Lemma. Define

$$y(t) = c_1 x_1(t) + c_2 x_2(t) + \cdots + c_n x_n(t), \ t \in I,$$

where $c_1, c_2,...,c_n$ are the solutions of the system given by (6). Then y is a solution of L(x) = 0 and in addition

$$y(a) = a_1, y'(a) = a_2, \cdots, y^{(n-1)}(a) = a_n. \tag{7}$$

From the uniqueness theorem, there is one and only one solution with these initial conditions. Hence y(t) = x(t) for t ∈ I. This completes the proof.

Method of Variation of Parameters

$$L(x) = d(t), \tag{8}$$

where L(x) is given by (2) or (4), is determined the moment we know x_h and x_p. It is therefore natural to know both a particular solution x_p of (8) as well as the general solution x_h of the homogeneous equation L(x) = 0. If L(x) = 0 is an equation with constant coefficients, the determination of the general solution is not difficult. Variation of parameter is a general method gives us a particular solution. The method of variation of parameters is also effective in dealing with equations with variable coefficients. To make the matter simple let us consider a second order equation

$$L(x(t)) = a_0(t)x''(t) + a_1(t)x'(t) + a_2(t)x(t) = d(t), \ a_0(t) \neq 0, \ t \in I, \tag{9}$$

where the functions $a_0, a_1, a_2, d : I \to \mathbb{R}$ are continuous. Let x_1 and x_2 be two linearly independent solutions of the homogeneous equation

$$a_0(t)x''(t) + a_1(t)x'(t) + a_2(t)x(t) = 0, \ a_0(t) \neq 0, \ t \in I. \tag{10}$$

Then, $c_1 x_1 + c_2 x_2$ is the general solution of (10), where c_1 and c_2 are arbitrary constants. The general solution of (9) is determined the moment we know a particular solution x_p of (9). We let the constants c_1, c_2 as parameters depending on t and determine x_p. In other words, we would like to find u_1 and u_2 on I such that

$$x_p(t) = u_1(t)x_1(t) + u_2(t)x_2(t), \ t \in I \tag{11}$$

satisfies (9).

In order to substitute x_p in (9), we need to calculate x_p' and x_p''. Now

$$x_p' = x_1' u_1 + x_2' u_2 + (x_1 u_1' + x_2 u_2').$$

We do not wish to end up with second order equations for u_1, u_2 and naturally we choose u_1 and u_2 to satisfy

$$x_1(t)u_1'(t) + x_2(t)u_2'(t) = 0 \tag{12}$$

Added to it, we already known how to solve first order equations. With (12) in hand we now have

$$x_p'(t) = x_1'(t)u_1(t) + x_2'(t)u_2(t). \tag{13}$$

Differentiation of (13) leads to

$$x_p'' = u_1'x_1' + u_1x_1'' + u_2'x_2' + u_2x_2''. \tag{14}$$

Now we substitute (11), (13) and (14) in (9) to get

$$[a_0(t)x_1''(t) + a_1(t)x_1'(t) + a_2(t)x_1(t)]u_1 + [a_0(t)x_2''(t) + a_1(t)x_2'(t) + a_2(t)x_2(t)]u_2 +$$
$$u_1'a_0(t)x_1' + u_2'a_0(t)x_2' = d(t),$$

and since x_1 and x_2 are solutions of (10), hence

$$x_1'u_1'(t) + x_2'u_2'(t) = \frac{d(t)}{a_0(t)}. \tag{15}$$

We solve for u_1' and u_2' from (12) and (15), to determine x_p. It is easy to see

$$u_1'(t) = \frac{-x_2(t)d(t)}{a_0(t)W[x_1(t),x_2(t)]}$$

$$u_2'(t) = \frac{x_1(t)d(t)}{a_0(t)W[x_1(t),x_2(t)]}$$

where $W[x_1(t), x_2(t)]$ is the Wronskian of the solutions x_1 and x_2. Thus, u_1 and u_2 are given by

$$\left. \begin{aligned} u_1(t) &= -\int \frac{x_2(t)d(t)}{a_0(t)W[x_1(t),x_2(t)]}dt \\ u_2(t) &= \int \frac{x_1(t)d(t)}{a_0(t)W[x_1(t),x_2(t)]}dt \end{aligned} \right\} \tag{16}$$

Now substituting the values of u_1 and u_2 in (11) we get a desired particular solution of the equation (9). Indeed

$$x_p(t) = u_1(t)x_1(t) + u_2(t)x_2(t), \quad t \in I$$

is completely known. To conclude, we have:

Theorem. Let the functions a_0, a_1, a_2 and d in (9) be continuous functions on I. Further assume that x_1 and x_2 are two linearly independent solutions of (10). Then, a particular solution x_p of the equation (9) is given by (11).

Theorem. The general solution $x(t)$ of the equation (9) on I is

$$x(t) = x_p(t) + x_h(t),$$

where x_p is a particular solution given by (11) and x_h is the general solution of L(x) = 0.

Also, we note that we have an explicit expression for x_p which was not so while proving Theorem. The following example is for illustration.

Example. Consider the equation

$$x'' - \frac{2}{t}x' + \frac{2}{t^2}x = t\sin t, \quad t \in [1, \infty).$$

Note that $x_1 = t$ and $x_2 = t^2$ are two linearly independent solutions of the homogeneous equation on $[1, \infty)$. Now

$$W[x_1(t), x_2(t)] = t^2.$$

Substituting the values of x_1, x_2, $W[x_1(t), x_2(t)]$, $d(t) = t\sin t$ and $a_0(t) \equiv 1$ in (16), we have

$$u_1(t) = t\cos t - \sin t$$
$$u_2(t) = \cos t$$

and the particular solution is $x_p(t) = -t\sin t$. Thus, the general solution is

$$x(t) = -t\sin t + c_1 t + c_2 t^2,$$

where c_1 and c_2 are arbitrary constants.

The method of variation of parameters has an extension to equations of order $n(n > 2)$ which we state in the form of a theorem, the proof of which has been omitted. Let us consider an equation of the n-th order

$$L(x(t)) = a_0(t)x^n(t) + a_1(t)x^{n-1}(t) + \cdots + a_n(t)x(t) = d(t), \quad t \in I. \tag{17}$$

Theorem. Let a_0, a_1,...,a_n, $d : I \to \mathbb{R}$ be continuous functions. Let

$$c_1 x_1 + c_2 x_2 + \cdots + c_n x_n$$

be the general solution of $L(x) = 0$. Then, a particular solution x_p of (17) is given by

$$x_p(t) = u_1(t)x_1(t) + u_2(t)x_2(t) + \cdots + u_n(t)x_n(t),$$

where u_1, u_2,..., u_n satisfy the equations

$$u_1'(t)x_1(t) + u_2'(t)x_2(t) + \cdots + u_n'(t)x_n(t) = 0$$
$$u_1'(t)x_1'(t) + u_2'(t)x_2'(t) + \cdots + u_n'(t)x_n'(t) = 0$$
$$\cdots\cdots\cdots\cdots\cdots\cdots\cdots\cdots\cdots\cdots\cdots\cdots$$
$$u_1'(t)x_1^{(n-2)}(t) + u_2'(t)x_2^{(n-2)}(t) + \cdots + u_n'(t)x_n^{(n-2)}(t) = 0$$
$$a_0(t)\left[u_1'(t)x_1^{(n-1)}(t) + u_2'(t)x_2^{(n-1)}(t) + \cdots + u_n'(t)x_n^{(n-1)}(t)\right] = d(t).$$

The proof of the Theorem is similar to the previous one with obvious modifications.

Two Useful Formulae

Two formulae proved below are interesting in themselves. They are also useful while studying boundary value problems of second order equations. Consider an equation

$$L(y) = a_0(t)y'' + a_1(t)y' + a_2(t)y = 0, \quad t \in I,$$

where $a_0, a_1, ..., a_n, d : I \to \mathbb{R}$ are continuous functions in addition $a_0(t)$ I$\neq$ 0 for t $\in$ I. Let u and v be any two twice differentiable functions on I. Consider

$$uL(v) - vL(u) = a_0(uv'' - vu'') + a_1(uv' - vu'). \tag{18}$$

The Wronskian of u and v is given by $W(u,v) = uv' - vu'$ which shows that

$$\frac{d}{dt}W(u,v) = uv'' - vu''.$$

Note that the coefficients of a_0 and a_1 in the relation (18) are $W'(u,v)$ and $W(u,v)$ respectively. Now we have

Theorem. If u and v are twice differential functions on I, then

$$uL(v) - vL(u) = a_0(t)\frac{d}{dt}W[u,v] + a_1(t)W[u,v], \tag{19}$$

where L(x) is given by (2). In particular, if L(u) = L(v) = 0 then W satisfies

$$a_0\frac{dW}{dt}[u,v] + a_1 W[u,v] = 0. \tag{20}$$

Theorem. (Able's Formula) If u and v are solutions of L(x) = 0 given by (2), then the Wronskian of u and v is given by

$$W[u,v] = k \exp\left[-\int \frac{a_1(t)}{a_0(t)}dt\right],$$

where k is a constant.

Proof: Since u and v are solutions of L(y) = 0, the Wronskian satisfies the first order equation (20) and Solving we get

$$W[u,v] = k \exp\left[-\int \frac{a_1(t)}{a_0(t)}dt\right] \tag{21}$$

where k is a constant.

The above two results are employed to obtain a particular solution of a non-homogeneous second order equation.

Example. Consider the general non-homogeneous initial value problem given by

$$L(y(t)) = d(t), \quad y(t_0) = y'(t_0) = 0, \quad t, t_0 \in I,$$

(22)

where $L(y)$ is as given in (9). Assume that x_1 and x_2 are two linearly independent solution of $L(y) = 0$. Let x denote a solution of $L(y) = d$. Replace u and v in (19) by x_1 and x to get

$$\frac{d}{dt} W[x_1, x] + \frac{a_1(t)}{a_0(t)} W[x_1, x] = x_1 \frac{d(t)}{a_0(t)}$$

(23)

which is a first order equation for W $[x_1, x]$. Hence

$$W[x_1, x] = \exp\left[-\int_{t_0}^{t} \frac{a_1(s)}{a_0(s)} ds\right] \int_{t_0}^{t} \frac{\exp\left[\int_{t_0}^{s} \frac{a_1(u)}{a_0(u)} du\right] x_1(s) ds}{a_0(s)} ds$$

(24)

While deriving (24) we have used the initial conditions $x(t_0) = x'(t_0) = 0$ in view of which W $[x_1(t_0), x(t_0)] = 0$. Now using the Able's formula, we get

$$x_1 x' - x x_1' = W[x_1, x_2] \int_{t_0}^{t} \frac{x_1(s) d(s)}{a_0(s) W[x_1(s), x_2(s)]} ds.$$

(25)

The equation (25) as well could have been derived with x_2 in place of x_1 in order to get

$$x_2 x' - x x_2' = W[x_1, x_2] \int_{t_0}^{t} \frac{x_2(s) d(s)}{a_0(s) W[x_1(s), x_2(s)]} ds.$$

(26)

From (25) and (26) one easily obtains

$$x(t) = \int_{t_0}^{t} \frac{\left[x_2(t) x_1(s) - x_2(s) x_1(t)\right] d(s)}{a_0(s) W[x_1(s), x_2(s)]} ds.$$

(27)

It is time for us to recall that a particular solution in the form of (27) has already been derived while discussing the method of variation of parameters.

Homogeneous Differential Equation

A differential equation can be homogeneous in either of two respects: the coefficients of the differential terms in the first order case could be homogeneous functions of the variables, or for the linear case of any order there could be no constant term.

Homogeneous Type of First-order Differential Equations

A first-order ordinary differential equation in the form:

$$M(x,y)dx + N(x,y)dy = 0$$

is a homogeneous type if both functions $M(x, y)$ and $N(x, y)$ are homogeneous functions of the same degree n. That is, multiplying each variable by a parameter λ, we find

$$M(\lambda x, \lambda y) = \lambda^n M(x,y) \text{ and } N(\lambda x, \lambda y) = \lambda^n N(x,y).$$

Thus,

$$\frac{M(\lambda x, \lambda y)}{N(\lambda x, \lambda y)} = \frac{M(x,y)}{N(x,y)}.$$

Solution Method

In the quotient $\dfrac{M(tx,ty)}{N(tx,ty)} = \dfrac{M(x,y)}{N(x,y)}$, we can let $t = 1/x$ to simplify this quotient to a function f of the single variable y/x:

$$\frac{M(x,y)}{N(x,y)} = \frac{M(tx,ty)}{N(tx,ty)} = \frac{M(1,y/x)}{N(1,y/x)} = f(y/x).$$

Introduce the change of variables $y = ux$; differentiate using the product rule:

$$\frac{d(ux)}{dx} = x\frac{du}{dx} + u\frac{dx}{dx} = x\frac{du}{dx} + u,$$

thus transforming the original differential equation into the separable form

$$x\frac{du}{dx} = f(u) - u;$$

this form can now be integrated directly.

The equations in this discussion are not to be used as formulary for solutions; they are shown just to demonstrate the method of solution.

Special Case

A first order differential equation of the form (a, b, c, e, f, g are all constants)

$$(ax + by + c)dx + (ex + fy + g)dy = 0$$

where $af \neq be$ can be transformed into a homogeneous type by a linear transformation of both variables (α and β are constants):

$$t = x + \alpha; z = y + \beta.$$

Homogeneous Linear Differential Equations

Definition. A linear differential equation is called homogeneous if the following condition is satisfied: If $\phi(x)$ is a solution, so is $c\phi(x)$, where c is an arbitrary (non-zero) constant. Note that in order for this condition to hold, each term in a linear differential equation of the dependent variable y must contain y or any derivative of y. A linear differential equation that fails this condition is called inhomogeneous.

A linear differential equation can be represented as a linear operator acting on $y(x)$ where x is usually the independent variable and y is the dependent variable. Therefore, the general form of a linear homogeneous differential equation is

$$L(y) = 0$$

where L is a differential operator, a sum of derivatives (defining the "0th derivative" as the original, non-differentiated function), each multiplied by a function f_i of x:

$$L = \sum_{i=0}^{n} f_i(x) \frac{d^i}{dx^i},$$

where f_i may be constants, but not all f_i may be zero.

For example, the following differential equation is homogeneous:

$$\sin(x)\frac{d\,y}{dx} + 4\frac{dy}{dx} + y = 0,$$

whereas the following two are inhomogeneous:

$$2x^2 \frac{d^2 y}{dx^2} + 4x\frac{dy}{dx} + y = \cos(x);$$

$$2x^2 \frac{d^2 y}{dx^2} - 3x\frac{dy}{dx} + y = 2.$$

It should be noted that the existence of a constant term is a sufficient condition for an equation to be inhomogeneous, as in the above example.

Homogeneous Linear Equations with Constant Coefficients

Homogeneous linear equations with constant coefficients is an important subclass of linear equations, the reason being that solvability of these equations reduces to he solvability algebraic equations. Now we attempt to obtain a general solution of a linear equation with constant coefficients. Let us start as usual with a simple second order equation, namely

$$L(y) = a_0 y'' + a_1 y' + a_2 y = 0, a_0 \neq 0. \tag{28}$$

Later we move onto a more general equation of order $n(n > 2)$

$$L(y) = a_0 y^{(n)} + a_1 y^{(n-1)} + \cdots + a_n y = 0$$

$$(29)$$

where $a_0, a_1,...,a_n$ are real constants and $a_0 \neq 0$.

Intuitively a look at the equation (28) or (29) tells us that if the derivatives of a function which are similar in form to the function itself then such a function might probably be a candidate to solve (28) or (29). Elementary calculus tell us that one such function is the exponential, namely e^{pt}, where p is a constant. If e^{pt} is a solution then,

$$L(e^{pt}) = a_0(e^{pt})'' + a_1(e^{pt})' + a_2(e^{pt}) = (a_0 p^2 + a_1 p + a_2)e^{pt}.$$

e^{pt} is a solution of (29) iff

$$L(e^{pt}) = (a_0 p^2 + a_1 p + a_2)e^{pt} = 0.$$

which means that e^{pt} is a solution of (29) iff p satisfies

$$a_0 p^2 + a_1 p + a_2 = 0.$$

$$(30)$$

Actually we have proved the following result:

Theorem. λ is a root of the quadratic equation (30) iff $e^{\lambda t}$ is a solution of (28).

If we note

$$L(e^{pt}) = (a_0 p^n + a_1 p^{n-1} + \cdots + a_n)e^{pt}$$

then the following result is immediate.

Theorem. λ is a root of the equation

$$a_0 p^n + a_1 p^{n-1} + \cdots + a_n = 0, \quad a_0 \neq 0$$

$$(31)$$

iff $e^{\lambda t}$ is a solution of the equation (29).

Definition. The equations (30) or (31) are called the characteristic equations for the linear differential equations (28) or (29) respectively. The corresponding polynomials are called characteristic polynomials.

In general, the characteristic equation (30) has two roots, say λ_1 and λ_2. By Theorem, $e^{\lambda_1 t}$ and $e^{\lambda_2 t}$ are two linearly independent solutions of (28) provided $\lambda_1 \neq \lambda_2$. Let us study the characteristic equation and its relationship with the general solution of (28).

Case 1: Let λ_1 and λ_2 be real distinct roots of (30). In this case $x_1(t) = e^{\lambda_1 t}$ and $x_2(t) = e^{\lambda_2 t}$ are two linearly independent solutions of (28) and the general solution x of (28) is given by $c_1 e^{\lambda_1 t} + c_2 e^{\lambda_2 t}$.

Case 2: When λ_1 and λ_2 are complex roots, from the theory of equations, it is well known that they are complex conjugates of each other i.e., they are of the form $\lambda_1 = a + ib$ and $\lambda_2 = a - ib$. The two solutions are

$$e^{\lambda_1 t} = e^{(a+ib)t} = e^{at}[\cos bt + i \sin bt],$$
$$e^{\lambda_2 t} = e^{(a-ib)t} = e^{at}[\cos bt - i \sin bt].$$

Now, if h is a complex valued solution of the equation (28), then

$$L[h(t)] = L[\operatorname{Re} h(t)] + iL[\operatorname{Im} h(t)], \ t \in I,$$

since L is a linear operator. This means that the real part and the imaginary part of a solution are also solutions of the equation (28). Thus

$$e^{at}\cos bt, e^{at}\sin bt$$

are two linearly independent solutions of (28), where a and b are the real and imaginary parts of the complex root respectively. The general solution is given by

$$e^{at}[c_1 \cos bt + c_2 \sin bt], \quad t \in I.$$

Case 3: When the roots of the characteristic equation (28) are equal, then the root is $\lambda_1 = -a_1/2a_0$. From Theorem 2.5.1, we do have a solution of (28) namely $e^{\lambda_1 t}$. To find a second solution two methods are described below, one of which is based on the method of variation of parameters.

Method 1: $x_1(t) = e^{\lambda_1 t}$ is a solution and so is $ce^{\lambda_1 t}$ where c is a constant. Now let us assume that

$$x_2(t) = u(t)e^{\lambda_1 t},$$

is yet another solution of (28) and then determine u. Let us recall here that actually the parameter c is being varied in this method and hence method is called Variation parameters.

Differentiating x_2 twice and substitution in (28) leads to

$$a_0 u'' + (2a_0\lambda_1 + a_1)u' + (a_0\lambda_1^2 + a_1\lambda_1 + a_2)u = 0.$$

Since $\lambda_1 = -a_1/2a_0$ the coefficients of u'and u are zero. So u satisfies the equation $u'' = 0$ whose general solution is

$$u(t) = c_1 + c_2(t), \quad t \in I,$$

where c_1 and c_2 are some constants or equivalently $(c_1 + c_2 t)e^{\lambda_1 t}$ is another solution of (28). It is easy to verify that

$$x_2(t) = te^{\lambda_1 t}$$

is a solution of (28) and x_1, x_2 are linearly independent.

Method 2: Recall

$$L(e^{\lambda t}) = (a_0\lambda^2 + a_1\lambda + a_2)e^{\lambda t} = p(\lambda)e^{\lambda t},$$

(32)

where $p(\lambda)$ denotes the characteristic polynomial of (28). From the theory of equations we know that if λ_1 is a repeated root of $p(\lambda) = 0$ then

$$p(\lambda_1) = 0 \text{ and } \left|\frac{\partial}{\partial\lambda}p(\lambda)\right|_{\lambda=\lambda_1} = 0.$$

(33)

Differentiating (32) partially with respect to λ, we end up with

$$\frac{\partial}{\partial\lambda}L(e^{\lambda t}) = \frac{\partial}{\partial\lambda}p(\lambda)e^{\lambda t} = \left[\frac{\partial}{\partial\lambda}p(\lambda) + tp(\lambda)\right]e^{\lambda t}.$$

But,

$$\frac{\partial}{\partial\lambda}L(e^{\lambda t}) = L\left(\frac{\partial}{\partial\lambda}e^{\lambda t}\right) = L(te^{\lambda t}).$$

Therefore,

$$L(te^{\lambda t}) = \left[\frac{\partial}{\partial\lambda}p(\lambda) + tp(\lambda)\right]e^{\lambda t}.$$

Substituting $\lambda = \lambda_1$ and using the relation in (33) we have $L\left(te^{\lambda_1 t}\right) = 0$ which clearly shows that $x_2(t) = te^{\lambda_1 t}$ is yet another solution of (29). Since x_1, x_2 are linearly independent, the general solution of (28) is given by

$$c_1 e^{\lambda_1 t} + c_2 te^{\lambda_1 t},$$

where λ_1 is the repeated root of characteristic equation (30).

Example. The characteristic equation of

$$x'' + x' - 6x = 0, \ t \in I,$$

is

$$p^2 + p - 6 = 0,$$

whose roots are p = −3 and p = 2. by case 1, e^{-3t}, e^{2t} are two linearly independent solutions and the general solution x is given by

$$x(t) = c_1 e^{-3t} + c_2 e^{2t}, \ t \in I.$$

Example. For

$$x'' - 6x' + 9x = 0, \ t \in I,$$

the characteristic equation is

$$p^2 - 6p + 9 = 0,$$

which has a repeated root p = 3. So (by case 2) e^{3t} and te^{3t} are two linearly independent solutions and the general solution x is

$$x(t) = c_1 e^{3t} + c_2 t e^{3t}, \quad t \in I.$$

The results which have been discussed above for a second order have an immediate generalization to a n-th order equation (29). The characteristic equation of (29) is given by

$$L(p) = a_0 p^n + a_1 p^{n-1} + \ldots + a_n = 0. \tag{34}$$

If p_1 is a real root of (34) then, $e^{p_1 t}$ is a solution of (29). If p_1 happens to be a complex root, the complex conjugate of p_1 i.e., $\bar{p}_1$ is also a root of (34). In this case

$$e^{at} \cos bt \ and \ e^{at} \sin bt$$

are two linearly independent solutions of (29), where a and b are the real and imaginary parts of p_1, respectively.

We now consider when roots of (34) have multiplicity(real or complex). There are two cases:

   (i) when a real root has a multiplicity m_1,

   (ii) when a complex root has a multiplicity m_1.

Case 1 : Let q be the real root of (34) with the multiplicity m_1. By induction we have m_1 linearly independent solutions of (29), namely

$$e^{qt}, te^{qt}, t^2 e^{qt}, \cdots, t^{m_1-1} e^{qt}.$$

Case 2 : Let s be a complex root of (34) with the multiplicity m_1. Let $s = s_1 + is_2$. Then, as in Case 1, we note that

$$e^{st}, te^{st}, \cdots, t^{m_1-1} e^{st}, \tag{35}$$

are m_1 linearly independent complex valued solutions of (29). For (29), the real and imaginary parts of each solution given in (35) is also a solutions of (29). So in this case $2m_1$ linearly independent solutions of (29) are given by

$$\left.\begin{array}{l} e^{s_1 t}\cos s_2 t,\ e^{s_1 t}\sin s_2 t \\ t e^{s_1 t}\cos s_2 t,\ t e^{s_1 t}\sin s_2 t \\ t^2 e^{s_1 t}\cos s_2 t,\ t^2 e^{s_1 t}\sin s_2 t \\ \cdots\cdots\cdots\cdots\cdots\cdots \\ t^{m_1-1} e^{s_1 t}\cos s_2 t,\ t^{m_1-1} e^{s_1 t}\sin s_2 t \end{array}\right\} \tag{36}$$

Thus, if all the roots of the characteristic equation (34) are known, no matter whether they are simple or multiple roots, there are n linearly independent solutions and the general solution of (29) is

$$c_1 x_1 + c_2 x_2 + \cdots + c_n x_n$$

where $x_1, x_2, \cdots, x_n$ are n linearly independent solutions and $c_1, c_2, \cdots, c_n$ are any constants. To summarize :

Theorem. Let $r_1, r_2, \cdots, r_s$, where $s \le n$ be the distinct roots of the characteristic equation (34) and suppose the root r_i has multiplicity m_i, $i = 1, 2, \cdots, s$, with

$$m_1 + m_2 + \cdots + m_s = n.$$

Then, the n functions

$$\left.\begin{array}{l} e^{r_1 t},\ t e^{r_1 t},\ \cdots,\ t^{m_1-1} e^{r_1 t} \\ e^{r_2 t},\ t e^{r_2 t},\ \cdots,\ t^{m_2-1} e^{r_2 t} \\ \cdots\cdots\cdots\cdots\cdots\cdots \\ e^{r_s t},\ t e^{r_s t},\ \cdots,\ t^{m_s-1} e^{r_s t} \end{array}\right\} \tag{37}$$

are the solutions of $L(x) = 0$ for $t \in I$

References

- Gershenfeld, Neil (1999), The Nature of Mathematical Modeling, Cambridge, UK.: Cambridge University Press, ISBN 978-0-521-57095-4

- Bôcher, Maxime (1901), "Certain cases in which the vanishing of the Wronskian is a sufficient condition for linear dependence", Transactions of the American Mathematical Society, Providence, R.I.: American Mathematical Society, 2 (2): 139–149, ISSN 0002-9947, JFM 32.0313.02, JSTOR 1986214, doi:10.2307/1986214

- Hartman, Philip (1964), Ordinary Differential Equations, New York: John Wiley & Sons, ISBN 978-0-89871-510-1, MR 0171038, Zbl 0125.32102

- Wolsson, Kenneth (1989a), "A condition equivalent to linear dependence for functions with vanishing Wronskian", Linear Algebra and its Applications, 116: 1–8, ISSN 0024-3795, MR 989712, Zbl 0671.15005, doi:10.1016/0024-3795(89)90393-5

- Robinson, James C. (2004), An Introduction to Ordinary Differential Equations, Cambridge, UK.: Cambridge University Press, ISBN 0-521-82650-0

- Wolsson, Kenneth (1989b), "Linear dependence of a function set of m variables with vanishing generalized Wronskians", Linear Algebra and its Applications, 117: 73–80, ISSN 0024-3795, MR 993032, Zbl 0724.15004, doi:10.1016/0024-3795(89)90548-X

Permissions

All chapters in this book are published with permission under the Creative Commons Attribution Share Alike License or equivalent. Every chapter published in this book has been scrutinized by our experts. Their significance has been extensively debated. The topics covered herein carry significant information for a comprehensive understanding. They may even be implemented as practical applications or may be referred to as a beginning point for further studies.

We would like to thank the editorial team for lending their expertise to make the book truly unique. They have played a crucial role in the development of this book. Without their invaluable contributions this book wouldn't have been possible. They have made vital efforts to compile up to date information on the varied aspects of this subject to make this book a valuable addition to the collection of many professionals and students.

This book was conceptualized with the vision of imparting up-to-date and integrated information in this field. To ensure the same, a matchless editorial board was set up. Every individual on the board went through rigorous rounds of assessment to prove their worth. After which they invested a large part of their time researching and compiling the most relevant data for our readers.

The editorial board has been involved in producing this book since its inception. They have spent rigorous hours researching and exploring the diverse topics which have resulted in the successful publishing of this book. They have passed on their knowledge of decades through this book. To expedite this challenging task, the publisher supported the team at every step. A small team of assistant editors was also appointed to further simplify the editing procedure and attain best results for the readers.

Apart from the editorial board, the designing team has also invested a significant amount of their time in understanding the subject and creating the most relevant covers. They scrutinized every image to scout for the most suitable representation of the subject and create an appropriate cover for the book.

The publishing team has been an ardent support to the editorial, designing and production team. Their endless efforts to recruit the best for this project, has resulted in the accomplishment of this book. They are a veteran in the field of academics and their pool of knowledge is as vast as their experience in printing. Their expertise and guidance has proved useful at every step. Their uncompromising quality standards have made this book an exceptional effort. Their encouragement from time to time has been an inspiration for everyone.

The publisher and the editorial board hope that this book will prove to be a valuable piece of knowledge for students, practitioners and scholars across the globe.

Index

www.ingramcontent.com/pod-product-compliance
Lightning Source LLC
Chambersburg PA
CBHW082048190326

41458CB00010B/3486